ENERGY SCIENCE, ENGINEERING AND TECHNOLOGY

BIOFUEL SUSTAINABILITY

RESEARCH AREAS AND KNOWLEDGE GAPS

ENERGY SCIENCE, ENGINEERING AND TECHNOLOGY

Additional books in this series can be found on Nova's website under the Series tab.

Additional E-books in this series can be found on Nova's website under the E-books tab.

AGRICULTURE ISSUES AND POLICIES

Additional books in this series can be found on Nova's website under the Series tab.

Additional E-books in this series can be found on Nova's website under the E-books tab.

ENERGY SCIENCE, ENGINEERING AND TECHNOLOGY

BIOFUEL SUSTAINABILITY

RESEARCH AREAS AND KNOWLEDGE GAPS

LAUREN S. GRAVER
AND
MATTHEW R. KRISS
EDITORS

Nova Science Publishers, Inc.
New York

Copyright © 2012 by Nova Science Publishers, Inc.

All rights reserved. No part of this book may be reproduced, stored in a retrieval system or transmitted in any form or by any means: electronic, electrostatic, magnetic, tape, mechanical photocopying, recording or otherwise without the written permission of the Publisher.

For permission to use material from this book please contact us:
Telephone 631-231-7269; Fax 631-231-8175
Web Site: http://www.novapublishers.com

NOTICE TO THE READER

The Publisher has taken reasonable care in the preparation of this book, but makes no expressed or implied warranty of any kind and assumes no responsibility for any errors or omissions. No liability is assumed for incidental or consequential damages in connection with or arising out of information contained in this book. The Publisher shall not be liable for any special, consequential, or exemplary damages resulting, in whole or in part, from the readers' use of, or reliance upon, this material. Any parts of this book based on government reports are so indicated and copyright is claimed for those parts to the extent applicable to compilations of such works.

Independent verification should be sought for any data, advice or recommendations contained in this book. In addition, no responsibility is assumed by the publisher for any injury and/or damage to persons or property arising from any methods, products, instructions, ideas or otherwise contained in this publication.

This publication is designed to provide accurate and authoritative information with regard to the subject matter covered herein. It is sold with the clear understanding that the Publisher is not engaged in rendering legal or any other professional services. If legal or any other expert assistance is required, the services of a competent person should be sought. FROM A DECLARATION OF PARTICIPANTS JOINTLY ADOPTED BY A COMMITTEE OF THE AMERICAN BAR ASSOCIATION AND A COMMITTEE OF PUBLISHERS.

Additional color graphics may be available in the e-book version of this book.

Library of Congress Cataloging-in-Publication Data

Biofuel sustainability : research areas and knowledge gaps / editors, Lauren S. Graver and Matthew R. Kriss.
    p. cm.
Includes index.
ISBN 978-1-62100-320-5 (hardcover)
1. Biomass energy. 2. Biomass energy--Research. I. Graver, Lauren S. II. Kriss, Matthew R.
TP339.B5385 2011
662'.88--dc23
              2011032585

Published by Nova Science Publishers, Inc. † New York

CONTENTS

Preface		**vii**
Chapter 1	Sustainability of Biofuels: Future Research Opportunities *USDA and United States Department of Energy*	**1**
Chapter 2	Measuring the Indirect Land-Use Change Associated with Increased Biofuel Feedstock Production *United States Department of Agriculture*	**51**
Index		**115**

PREFACE

Legislative mandates and incentives, volatility in oil prices and new research and technological advances are driving the expectation of major increases in the production of biofuels from cellulosic biomass. The term "sustainability" has been defined as the meeting of needs of present and future generations. Sustainable biofuel production is economically competitive, conserves the natural resource base and ensures social well-being. This book explores critical research areas and knowledge gaps relevant to the environmental, economic and social dimensions of biofuel sustainability. It also underscores the critical need for a common socioeconomic framework to develop a systems-level understanding for how these dimensions interact across different spatial scales, from the small plot or farm to regional to very large scales such as political, national and global scales.

Chapter 1- Legislative mandates and incentives, volatility in oil prices, and new research and technological advances are driving the expectation of major increases in the production of biofuels from cellulosic biomass. To assess the current state of the science underlying the sustainability of an emergent cellulosic biofuel sector and to identify further research needs, the U.S. Department of Agriculture's Research, Education, and Economics mission area and the U.S. Department of Energy Office of Science cosponsored a workshop on October 28–29, 2008. Although the term "sustainability" has been defined in many ways, common to these definitions is the theme of meeting the needs of present and future generations. Sustainable bio-fuel production is economically competitive, conserves the natural resource base, and ensures social well-being. This report summarizes critical research areas and knowledge gaps relevant to the environmental, economic, and social dimensions of biofuel sustainability. It also underscores the critical need for a common socioecological framework to develop a systems-level understanding for how these dimensions interact across different spatial scales—from the small plot or farm to regional to very large scales such as political, national, and global scales. In addition to forging a responsible path for implementing cellulosic biofuels, much of what can be discovered about biofuel sustainability will provide important insights into successful future agricultural and forest production—the dependable and abundant supply of food, fiber, and feed.

Chapter 2- In recent years, concerns have been raised about potential domestic and international land-use changes that might be associated with scaling up biofuel feedstock production. Increased competition for productive land and the resulting shifts in land use to produce food, feed, fiber, and fuel have potential impacts on greenhouse gas emissions, biodiversity, and water quality. The House Report 111-181 accompanying H.R. 2997, the 2010 Agriculture, Rural Development, Food and Drug Administration, and Related Agencies Appropriations Bill, requested the USDA's Economic Research Service (ERS) in conjunction with the Office of the Chief Economist, to conduct a study of land-use changes for renewable fuels and feedstocks used to produce them. This report is a response to that request, and it summarizes the current state of knowledge of the drivers of land-use change and the role of biofuel production in affecting land-use change. The analytical frameworks that have been used to address land-use impacts of increased biofuels production are presented, and the estimation task is explored in more detail. The objective was to survey the literature in a neutral, objective way. There was no intention to suggest that USDA does or does not agree with any assumptions or model results.

In: Biofuel Sustainability: Research Areas...   ISBN: 978-1-62100-320-5
Editors: L. S. Graver, M. R. Kriss   © 2012 Nova Science Publishers, Inc.

Chapter 1

SUSTAINABILITY OF BIOFUELS: FUTURE RESEARCH OPPORTUNITIES[*]

USDA and United States Department of Energy

Convened by
U.S. Department of Energy
Office of Science
Office of Biological and
Environmental Research

U.S. Department of Agriculture
Research, Education,
and Economics

Organizers

John C. Houghton, Ph.D.
Office of Science
U.S. Department of Energy

---

[*] This is an edited, reformatted and augmented version of the USDA and United States Department of Energy publication, DOE/SC-0114, dated on October 2008.

Jeffrey Steiner, Ph.D. Agricultural Research Service
U.S. Department of Agriculture
Elizabeth White, MBA, MPH
Office of Science
U.S. Department of Energy

Jeffrey Steiner, Ph.D.
Agricultural Research Service
U.S. Department of Agriculture

Marilyn Buford, Ph.D.
U.S. Forest Service
U.S. Department of Agriculture

Patricia Hippie, Ph.D.
Cooperative State Research,
Education, and Extension Service
U.S. Department of Agriculture

Robbin Shoemaker, Ph.D.
Economic Research Service
U.S. Department of Agriculture

As part of the U.S. Department of Energy's (DOE) Office of Science, the Office of Biological and Environmental Research supports fundamental research and technology development aimed at achieving predictive, systems-level understanding of complex biological and environmental systems to advance DOE missions in energy, climate, and environment.

The U.S. Department of Agriculture's (USDA) mission area in Research, Education, and Economics is dedicated to the creation of a safe, sustainable, and competitive U.S. food and fiber system, as well as strong communities, families, and youth through integrated research, analysis, and education.

Suggested citation for this report: U.S. DOE and USDA, 2009. *Sustainability of Riofueis:* Future *Research Opportunities; Report from the October 2008 Workshop,* DOE/SC-0114, U.S. Department of Energy Office of Science and U.S. Department of Agriculture (http://genomicsgtLenergy.gov/biofuels/sustainabilityand http://www.ree.usda.govi).

Sources for cover images: Poplar researcher image courtesy of Oak Ridge National Laboratory. Poplar forest and ethanol plant images courtesy of the

Sustainability of Biofuels: Future Research Opportunities 3

U.S. DOE National Renewable Energy Laboratory. Aerial crop and economic researcher images courtesy of USDA Agricultural Research Service. Soil and farmer outreach images courtesy of USDA Natural Resources Conservation Service.

EXECUTIVE SUMMARY

Legislative mandates and incentives, volatility in oil prices, and new research and technological advances are driving the expectation of major increases in the production of biofuels from cellulosic biomass. To assess the current state of the science underlying the sustainability of an emergent cellulosic biofuel sector and to identify further research needs, the U.S. Department of Agriculture's Research, Education, and Economics mission area and the U.S. Department of Energy Office of Science cosponsored a workshop on October 28–29, 2008. Although the term "sustainability" has been defined in many ways, common to these definitions is the theme of meeting the needs of present and future generations. Sustainable bio-fuel production is economically competitive, conserves the natural resource base, and ensures social well-being. This report summarizes critical research areas and knowledge gaps relevant to the environmental, economic, and social dimensions of biofuel sustainability. It also underscores the critical need for a common socioecological framework to develop a systems-level understanding for how these dimensions interact across different spatial scales—from the small plot or farm to regional to very large scales such as political, national, and global scales. In addition to forging a responsible path for implementing cellulosic biofuels, much of what can be discovered about biofuel sustainability will provide important insights into successful future agricultural and forest production—the dependable and abundant supply of food, fiber, and feed.

Environmental Dimensions of Biofuel Sustainability

The four dimensions of environmental sustainability research in this report are (1) soil resources and greenhouse gas emissions; (2) water quality, demand, and supply; (3) biodiversity and ecosystem services; and (4) integrated landscape ecology and feedstock production analysis.

4   USDA and United States Department of Energy

High-yielding feedstock production systems require soil resources that allow sufficient root penetration and provide adequate nutrient and water supplies throughout the growing season. Soils also can play a role in mediating climate change by storing carbon and providing habitats for microbial communities that influence greenhouse gas emissions or help promote efficient production of plant feedstocks that can be converted into biofuels. Some key research opportunities include using advanced microbial genomics to enhance soil fertility and reduce greenhouse gas emissions, characterizing and modeling soil carbon and nitrogen cycling processes for different biofuel feedstock production systems, developing improved biofuel feedstock production systems, and predicting how soil and plant processes will respond to climate change. Also needed is the development of field-deployable instrumentation to quantify nitrous oxide and methane fluxes.

Up to now, most research on cellulosic feedstocks has focused on optimizing growth conditions and feedstock productivity. Research is needed to determine the impacts of biomass feedstock production on water quality and availability within different ecoregions where biofuels will be produced. The impacts of biofuel production on water quality, demand, and supply will vary considerably by region, depending upon competition for water supply, the type of biomass feedstock and how it is managed, characteristics of the land, and the local climate, including potential future changes. Research will help determine the potential impacts of converting existing managed and natural landscapes to bioenergy feedstock production and developing regional assessments of water requirements and impacts for a wide range of feedstocks and management practices.

Biofuel production systems will be a part of larger landscapes that provide a variety of ecosystem services important to society and the environment. In addition to crop productivity, control of greenhouse gases, and reduction of water contamination, biofuels could increase biodiversity and supply critical habitats for beneficial organisms. As well as gaining a more comprehensive understanding of the effects of different biofuel production systems on the provision and regulation of ecosystem services, analysis frameworks are needed for modeling and bundling some of the most highly valued services for optimal multifunctional benefits.

Landscapes used for biofuel production will be characterized by complex interactions with a large number of ecosystem components that act together in important but as-yet incompletely understood ways. Information is needed at intermediate watershed and regional scales of resolution, and research will be critical to understand and improve models of ecosystem biophysical properties and their interactions and integration with economic and other human behaviors.

Economic Dimensions of Biofuel Sustainability

An active area of economic research is determining the mix of cellulosic feedstocks likely to be competitive in different regions, the spatial pattern of land-use changes that the use of these feedstocks will induce, and implications for food prices. Although existing economic models can predict aggregate land-use change, more precise estimates—particularly at smaller scales—will require additional research. Further research on indirect land-use effects will help resolve implications of increased production of feedstocks. Life cycle analyses should be improved to capture more accurately a comprehensive real-world representation of the ancillary effects of management scenarios. In addition, economic analyses are needed to examine the costs and benefits of policies to achieve biofuel production goals and to determine possible unintended consequences.

Social Dimensions of Biofuel Sustainability

The social implications of the emergence of cellulosic biofuels represent some of the most pressing and challenging sustainability issues. Research to understand how stakeholders may respond based on their values, choices, behaviors, and reactions will be critical to the development of a biofuel sector. Careful consideration must be given to social structures and policies that can promote or inhibit development of expanded biofuel production. As with biophysical considerations, adequate analyses will be necessary to understand how social processes function at multiple scales and with complexity—from individual farms and forests to whole communities and regional ecosystems— so science can inform decision making and design at local, regional, national, and global levels. Research is needed to identify feedstock production systems, biorefining processes, and enterprise structures that fit the needs and values of different communities and to optimize benefits for biomass producers, biorefiners, and encompassing communities by improving local conditions and reducing undesirable consequences.

Research to understand stakeholder needs and motivations will help define preferred societal outcomes. A diverse portfolio of decision aids, education, communication tools, and outreach and extension activities will be needed to enable stakeholders to make decisions based on information supported by environmental, economic, and social science research.

1. INTRODUCTION

The U.S. Department of Agriculture (USDA) and the U.S. Department of Energy (DOE) held a Sustainability of Biofuels Workshop on October 28–29, 2008, in Bethesda, Maryland. Its purpose was to assess the current state of the science and to identify further research needs in the effort to develop a sustainable biofuel economy.

The workshop was jointly hosted by USDA Under Secretary for Research, Education, and Economics Gale Buchanan and DOE Under Secretary for Science Raymond L. Orbach. DOE and USDA have the joint goal of informing the debate surrounding the sustainability of biofuels by providing sound science through strategic investment in research programs. This report describes issues addressed at the workshop and identifies critical areas and knowledge gaps that can be advanced by further sustain-ability research. The report also summarizes research opportunities identified by workshop participants and is organized around themes based on the three dimensions of sustainability:

- Environmental
- Economic
- Social

Although "sustainability" has been defined many different ways, underlying all these definitions is the common theme of meeting the needs of present and future generations while conserving the natural resource base and ensuring social and environmental well-being. The sustainability of biofuels (or any product) spans environmental, economic, and social dimensions that interconnect.

One strong message from the discussions was the need for a common socioecological framework for the study of sustainability and for a systems approach across scales. Successful biofuel development will depend on understanding the complex, integrated nature of sustainability. This knowledge must be used to build a new biofuel sector by considering production costs and environmental outcomes, as well as local, regional, national, and global needs.

In addition to forging a path for implementing cellulosic biofuels "the right way," much of what can be discovered about biofuel sustainability will provide important insights into the successful future production of food, fiber, and feed. Success will require an integrated, holistic approach to research and implementation that cuts across the environmental, economic, and social aspects of biofuel sustainability.

No single feedstock type or land-management practice will work for all locations. To understand the kinds of feedstocks and management regimes that would be best suited for different landscapes, it is necessary to envision the complete system—from production, management, and processing to ecosystem services, and from economic outputs to infrastructure and resource requirements for local production of different feedstocks. Research needs range from genomic tools that target soil microbial communities to those that measure the state of natural resources under different production scenarios and tools to understand the social and economic implications of decisions that influence the selection and implementation of biofuel feedstocks.

Research will be needed to address ways to determine impacts at diverse scales from the molecular level to entire regions of the country. Tools will be needed for integrating and extrapolating information derived from all operational scales. For example, just as more needs to be understood about how water use is regulated at a fundamental molecular level by a particular feedstock, more also needs to be understood regarding the impacts of how managed inputs such as fertilizer applications may influence nitrate loading from edges of fields, the quality of water at a small watershed scale, and, to a greater extent, the impact of summer hypoxia events in large receiving water bodies.

Biofuel production has regional and global implications on food, fiber, and feed production and for the provision of ecosystem services such as soil and water quality and biological diversity. Choices of biomass feedstocks, cultivation and harvesting practices, and technological changes could have a variety of potential impacts. Research is needed to develop models to predict the most significant outcomes that could ripple through interconnected ecological and societal systems. The illustration (see sidebar, Socioecological Framework for Biofuel Systems, p. 7) describes a cyclic framework in which cropping systems provide ecosystem services, which are valued by society through economics and other social systems and consequently affect management choices for the cropping systems.

Growth of the biofuel sector will take place in a dynamic fashion influenced by changing environmental, economic, societal, and technological

8   USDA and United States Department of Energy

factors. For example, the cost that society is willing to incur for contaminated runoff, loss of soil fertility, erosion, and disruption of wildlife habitat may change. Innovations from material science and biotechnology are likely to lead to major advances in energy feedstocks and fuel products. Population growth, climate change, globalization or localization of energy and other markets, and changes in the way energy is generated (distributed versus centralized) also will have significant impacts. Research is needed to understand how this dynamic environment will impact future opportunities and needs for biofuel use and development.

Successful expansion of cellulosic biofuels requires new transformational technologies that address challenges to sustainability such as reliability of abundant feedstock supplies; land-use change and competition; cost reductions for growing, harvesting, and transporting feedstocks; the efficiency of feedstock conversion; and the production and utilization of conversion by-products. Research is needed to ensure the development and availability of integrated production systems that are flexible in the face of evolving innovations, developing knowledge, and identification of best practices.

Research is needed to develop decision-support tools that help decision and policy makers weigh alternatives, anticipate likely outcomes, identify important factors and tradeoffs, and quantify uncertainties of decisions at the farm, forest, community, regional, national, and global scales. For example, a tool that helps select the best location to site a biofuel production facility might incorporate information on feedstock availability, type, and growth rates; infrastructure; capital and labor markets; and tax structure. Similarly, science-based performance measurements for farm and forest management and biorefinery operation are needed to compare outcomes at all levels from local to global.

Socioecological Framework for Biofuel Systems

Currently, most analyses and ecological modeling of biofuels focus on the biophysical aspects of the cropping systems and conversion technologies used for biofuel production. Developing sustainable biofuels, however, will require understanding how biofuel production will influence and be affected by interconnected social systems, including economics, and ecosystems. The figure above shows how managed and unmanaged disturbances shaped by human behaviors can impact cropping systems and associated ecosystem services, which, in turn, feed back into the social system affecting human decisions, behaviors, and outcomes. Research that

Sustainability of Biofuels: Future Research Opportunities

provides a comprehensive view of the interactions among bioenergy cropping systems, social systems, and ecosystem services is needed to develop science-based informational resources that can support decision making at local to national and global levels.

Socioecological Framework for Biofuel Systems

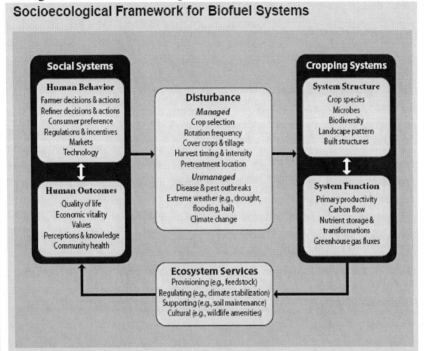

After Robertson et al. (in prep.) after Collins et al. 2007.

Source: After Robertson et al. (in prep.) after Collins et al. 2007. *Integrated Science for Society and the Environment: A Strategic Research Initiative.* Publication #23 of the U.S. Long-Term Ecological Research Network (LTER), LTER Network Office, Albuquerque, New Mexico.

1.1. Driving Forces for Biofuel Development

Countries and companies across the globe are investing extensively in biofuel development programs, motivated by concerns and opportunities related to global climate change, energy security, and economic development. Many countries have ambitious biofuel targets or mandates.

Grain-based biofuels already provide some nations with a renewable energy resource that has produced new jobs and economic development opportunities. Agricultural and forest producers, biorefiners, and policy makers anticipate that cellulosic biofuels have the potential to achieve ambitious national goals for biofuel production. Sound science, technology, economics, and policy development will be needed to ensure the sustainable production of cellulosic biomass, including intensification and potential expansion of agricultural and silvicultural practices to meet the demand for biofuels, conserve or enhance natural resources, and benefit farm and forest economies and rural communities.

In the United States, the diverse goals for accelerated production of biofuels from agricultural and forest resources are reflected in a series of recent U.S. policies: the Biomass Research and Development Act of 2000, the Energy Policy Act of 2005, the 2002 and 2008 Farm Bills, and the Energy Independence and Security Act (EISA) of 2007. As part of EISA, the Renewable Fuel Standard (RFS) mandates that 36 billion gallons of biofuels are to be produced annually by 2022, of which 16 billion gallons are expected to come from cellulosic feedstocks that will need to be produced from working lands on a large scale (see sidebar, Land Requirements for Biofuel Production, p. 9). EISA also includes a variety of incentives for the demonstration and deployment of biofuel production technologies, including biorefinery plant construction and operation, and describes requirements and subsidies for the use of biologically derived ethanol in gasoline blends. In addition to provisions for biofuel production, EISA recognizes the importance of biofuel sustainability by mandating a life cycle analysis for biofuels every 2 years and the development of sustainability criteria and indicators. The Food, Conservation, and Energy Act of 2008 (P.L. 110- 246, 2008 Farm Bill) includes incentives and programs for accelerating cellulosic feedstock production and cellulosic biofuel production and refining.

With the recent surge of national and political support for the large-scale development of bioenergy alternatives to fossil fuels, some of the most important issues arising from the potential paradigm shift for bioenergy production are the environmental, economic, and social implications. Subsequent chapters in this report summarize output from workshop participants and identify key challenges, knowledge gaps, and research opportunities specific to environmental, economic, and social dimensions of biofuel sustainability.

1.2. Sustainability and the Emerging Cellulosic Biofuel Industry

The emerging cellulosic biofuel industry—if driven by science-based strategies that conserve or enhance the natural resource base, increase economic viability, and build societal acceptance—offers the potential for new sustainable outcomes that have not been achieved with existing grain- and petroleum-based systems alone. Agricultural and forest landscapes that will be called upon for this next generation of biofuels need to be viewed as sources of multiple benefits, including biofuels (Jordan et al. 2007). Such benefits include carbon sequestration, conserved and enhanced soil productivity, reduced greenhouse gas emissions, and increased economic development of rural communities.

Land Requirements for Biofuel Production

Land area in the United States is about 2.3 billion acres. Around 1 billion acres are used for agricultural purposes (including grasslands, pasture, and croplands); 650 million acres are forest-use lands; and the remaining portion is devoted to parks and wildlife areas, urban areas, and other miscellaneous uses (Lubowski et al. 2006). About 340 million acres of agricultural lands are active cropland with corn, soybeans, and wheat representing around two-thirds of this area. Even with the recent growth in corn ethanol production, only about 18% of the grain harvested from 87 million acres of corn in the United States was used for ethanol production in 2007, while more than half of harvested corn grain was used for animal feed (USDA/ NASS 2008; USDA ERS 2008).

The 2007 Energy Independence and Security Act Renewable Fuels Standard has mandated the production of 16 billion gallons of cellulosic biofuels by 2022. To meet this target, cellulosic biomass will need to be harvested from America's working lands on a large scale. By one estimate, 16 to 19 million acres of energy crops are needed (Biomass Research and Development Board 2008).

Feedstock production systems designed around improved crops and practices could require less fertilizer and perhaps less water, trap nitrogen and phosphorus that otherwise would be transported to groundwater and streams, and accumulate carbon in both roots and soil organic matter. Effectively managed, these feedstock systems additionally could enhance ecosystem services such as natural insect and disease pest suppression, water-quality protection, and cultural and wildlife amenities. However, these potential benefits are not guaranteed. Uninformed or short-term decisions about how, when, and where cellulosic feedstocks and biofuels are produced could limit progress toward a sustainable bioenergy future (Robertson et al. 2008).

2. ENVIRONMENTAL DIMENSIONS OF BIOFUEL SUSTAINABILITY

Ensuring that the emerging cellulosic biofuel industry is sustainable requires careful consideration of environmental dimensions. Soil quality, which is determined by a complex collection of biogeochemical processes, is important to protect for both current and future crops and thus is a key area for research. Similarly, an accurate accounting of greenhouse gas emissions associated with different types of feedstock production is necessary to understand the impact of cellulosic biofuel production on ecosystem health and the quality of natural resources. In addition, the increasing pressures on water supplies nationwide are expected to continue, requiring research to minimize water use by biomass crops, as well as nutrient and other contaminant runoff. Another important research area involves understanding the role of biodiversity in maintaining ecosystem services and developing necessary strategies to ensure that as the production of biofuels increases, adequate supplies of other needed agricultural and forest-based goods are produced. Finally, fully understanding the potential impacts of biofuel production on landscape ecology and systems interactions requires expansion of field experiments and modeling studies beyond the small-plot and field scales to regional scales with appropriate validation and interpretation using real-world biophysical, economic, and social conditions. Each of these areas of environmental sustainability research is discussed more thoroughly in the following sections of this chapter. Using genomics and systems biology approaches to improve potential bioenergy crops and obtain a mechanistic understanding of the biological processes underlying bioenergy feedstock development is important, but it was not a focus for this workshop. Biological feedstock development research topics are presented in the report *Breaking the Biological Barriers to Cellulosic Ethanol*, based on a workshop convened by the U.S. Department of Energy in late 2005.

2.1. Soil Resources and Greenhouse Gas Emissions

Soil is the foundation of plant production. It determines the kinds of plants that can be grown; the need for water, organic matter, and nutrient amendments; and the outcomes that result. High-yielding production systems can occur only when the soil provides an adequate water supply throughout the growing season, allows roots to penetrate the soil profile to use nutrients and water, and presents minimal limitations to plant growth and development. Soils also can play a role in mitigating climate change by enhancing carbon storage and by providing habitat to support microbes that generate or consume the greenhouse gases methane and nitrous oxide.

When feedstock production systems are not managed to protect soil resources, degradation can occur, resulting in soil loss due to water and wind erosion, reduced water and nutrient availability, and deterioration of soil structure that can limit rooting depth, aeration, and water movement. Additionally, stored carbon can be released back to the atmosphere through natural processes and production management practices. As biofuel feedstock production expands, research is needed to determine how crop and forest production systems might be made more efficient while at the same time maintaining or enhancing soil productivity.

Many cellulosic feedstocks are perennial (either as a monoculture or polyculture), and, consequently, roots are always present to help reduce soil erosion and retain nutrients. Research is needed to determine how perennial plants, their root structure, and associated microbial communities impact belowground carbon allocation and greenhouse gas production. Research focused on improving quantification of soil carbon and nutrient cycling processes, including the movement of carbon through short- and long-lived soil carbon pools, is needed to better understand and manage systems to conserve soil carbon. Also important is research to better define relationships among soil carbon storage and the fluxes of non-CO_2 greenhouse gases under perennial crops.

Climate variability lies at the crux of optimizing feedstock production systems. To effectively manage plant productivity and soil carbon processes, research is needed to better understand potential changes in precipitation and temperature patterns, atmospheric carbon dioxide concentrations, nutrient availability, and resistance to disturbances.

Understanding Interactions among Soil, Microbial, and Plant Processes

Plants are complex systems in which biogeochemical interactions occur among the carbon, water, and nitrogen cycles and the microbes that live around the roots, on leaves, and as endophytes living inside the plant. Atmospheric CO_2 is taken up by plants through photosynthesis. Although some of this carbon is respired back to the atmosphere, a portion is incorporated into plant material distributed above and below ground. Nonharvested plant material becomes soil organic matter (SOM), which is linked to the water cycle as it impacts infiltration rates and the soil's water-holding capacity. These two cycles, in turn, are linked to cycles of nitrogen and other nutrients crucial for plant growth and development.

Microbes in the plant environment can fix nitrogen, mineralize nutrients from decaying organic matter, scavenge phosphorus, produce plant growth promoters, aid soil structure, and protect against disease agents. These functions help improve biofuel feedstock production efficiency while also ensuring sustainability of the soil resource. Control of soil microbes could play a beneficial role in increasing the production system efficiency of biofuel crops. On average, a third of the nitrogen used by sugarcane can be acquired via a nitrogen-fixing system (Boddey et al. 2003; Polidoro 2001; de Resende et al. 2006). Some of these bacteria also produce hormones that stimulate plant growth (Baldani and Baldani 2005). Research to understand and apply these and other strategies could result in increased production efficiency of dedicated energy crops with reduced dependence on synthetic fertilizers.

Enhancing Carbon Storage

Research on biogeochemical processes that influence carbon storage and fluxes in soils may also lead to decreased emissions of CO_2. Carbon storage is controlled by the soil environment and the quality of the organic matter in which the carbon resides. Maintenance of optimal soil water and temperature regimes results from soil management strategies that protect the soil microenvironment and promote an environment conducive to beneficial microbial activity. The addition of organic matter and maintenance of soil cover can improve soil quality by building SOM and can thereby lead to higher plant productivity and other environmental benefits.

Minimizing Net Non-CO_2 Greenhouse Gas Emissions

In addition to CO_2 emissions, another important area of research is to understand the non-CO_2 greenhouse gases methane and nitrous oxide that are associated with biofuel production. Although CO_2 is the most abundant greenhouse gas, both methane and nitrous oxide are more potent, with global-warming potentials much higher than that of CO_2. In general, nitrous oxide releases increase with the addition of excess nitrogen fertilizer and thus can either decrease or eliminate the greenhouse gas benefits of a biofuel operation.

Besides greenhouse gas emissions, various levels of air pollutants—carbon monoxide, volatile organic compounds, fine particles, and sulfur oxides—are released on end use, depending on the type of biofuel blend used for combustion. In general, air pollutant emissions from biofuel combustion tend to be lower than those from the combustion of petroleum-based fuels. Although these pollutants have important air quality and human health impacts, they were not a key focus for this workshop.

Measuring Flows and Stocks of Greenhouse Gases

New monitoring and instrumentation will be necessary to measure, predict, and manage the flows and stocks of CO_2, nitrous oxide, and methane in biofuel systems. Net soil CO_2 release is commonly inferred from soil carbon change, which means that carbon stored in soil will need to be estimated carefully, including its form and persistence.

Harvesting Biomass while Maintaining Site Productivity

Research on the influence of biofuel crops and management strategies on soil fertility will help improve plant productivity. Expanded crop yields require the efficient use of carbon, water, nitrogen, and other nutrients by the plant, which, in turn, requires a soil resource capable of supplying water and nutrients to meet plant requirements.

Soil protection measures should be integrated into biomass production methodologies. Research is needed that includes long-term soil-quality monitoring to help assess changes in physical, chemical, microbial, and other biological properties, thereby providing critical information for designing management systems to support bioenergy production (Wilhelm et al. 2007), habitat restoration, and the reduction of wildfire risk.

There also is substantial interest in using accumulating forest biomass for biofuels. In the western United States, biomass buildup as a result of fire suppression and insect and disease outbreaks on federal lands is a primary motivator for removal. Converting this buildup into cellulosic biofuel could help suppress unmanaged wildfires, improve stand health, and meet cellulosic feedstock needs.

Research is needed to develop appropriate harvest and collection systems to protect site hydrology and soil structure and productivity.

Reducing Net Greenhouse Gas Emissions from Cultivation and Harvest Practices

Research is needed to determine impacts of the cultivation and harvest of cellulosic feedstocks on greenhouse gas emissions in order to improve those practices. Conservation tillage farming can reduce erosion, and, when carried out over long periods of time, it can improve soil carbon content. Net greenhouse gas emissions can be reduced by conservation tillage even when combined with the application of nitrogen fertilizer (Archer and Halvorson 2009, in review). More research on options to integrate bioenergy crop production with existing row crops could reduce the greenhouse gas footprint for traditional agriculture by decreasing the need for fertilizers and minimizing carbon and nitrous oxide emissions from soils.

Research Opportunities

- Improve biofuel crop performance and soil fertility by understanding and manipulating microbial communities.
  - Use genomics and other advanced methods to better characterize the function of microbial communities in plant-soil systems, including the rhizosphere, foliar, and endophytic microbes involved in carbon and nitrogen cycling, disease suppression, and other services.
  - Develop improved understanding of the biotic and physic-chemical factors that control the distribution, abundance, and effectiveness of these microbes, including interactions with organisms in other trophic levels and how these affect the production of biofuel feedstocks and other agricultural and forest products.
- Predict and manipulate soil carbon cycling and sequestration.
  - Investigate differences among candidate biofuel management systems with respect to belowground carbon cycling and potential rates of carbon sequestration.
  - Characterize the biochemical nature and recalcitrance of sequestered carbon and its importance for soil structure and nutrient and water availability, while also developing an improved understanding of the biotic and physicochemical factors

Sustainability of Biofuels: Future Research Opportunities   17

that control the persistence of sequestered carbon and how these are influenced by management.

- Evaluate and develop improved biofuel feedstock production systems.
  - Build improved quantitative models of carbon, nitrogen, and water cycles in biofuel feedstock production systems to predict productivity and environmental outcomes from field to landscape scales. Create a means to link biophysical models to land-use, economic, and other socioecological models in order to simultaneously forecast the outcomes of alternative policy and land-use decisions in different biophysical, socioeconomic, and soil domains.
  - Identify and understand response thresholds, such as the reaction of soil carbon or microbial communities, to differences in the rate of agricultural or forest residue removal or to differences in the intensity of management through fertilizers or other inputs. Identify other biology-based management strategies that reduce the need for agricultural inputs.
  - Determine how to identify resistance and resilience of different systems as challenged by biogeochemical or technological change, as well as the mechanisms responsible for these differences.
- Predict responses of soil and biomass productivity to climate change.
  - Build on existing experiments and infrastructure to predict how agricultural and forest ecosystems will respond to climate change and changes in atmospheric chemistry. Use multiscale infrastructure from fields to farms to watersheds and regions so precipitation, temperature, and other environmental factors can be manipulated to understand their interactions and significant impacts on systems.
- Understand how soil microbial populations and activity influence methane and nitrous oxide consumption and fluxes to minimize emissions.
  - Use genomic and other advanced approaches to characterize how soil microbial communities respond to management and are responsible for nitrous oxide production and methane consumption.
  - Determine how changing biotic, physical, and chemical factors control soil microbial distribution, abundance, and capacity to

produce and consume trace gas and determine how populations can be controlled for more sustainable outcomes.

- Model nitrous oxide and methane fluxes to identify strategies that will reduce emissions from cropping systems.
- Conduct long-term field experiments to characterize fluxes in soil carbon during the establishment and production phases of cropping systems and in response to changing annual and longer-term environmental conditions. Quantify long-term trends in nitrous oxide and methane fluxes in cellulosic biofuel systems and the environmental and management factors that regulate fluxes at different temporal and spatial scales across U.S. ecoregions.
- Refine and validate mechanistic models of nitrous oxide and methane fluxes for different biofuel cropping systems in an appropriate variety of climate and soil domains. Incorporate soil management and best management practices into landscape-level models to allow the prediction of fluxes with land-use change. Develop and test decision support tools that can be used by producers and decision makers to design high-mitigation biofuel systems.
- Develop field-deployable instrumentation for quantifying *in situ* nitrous oxide and methane fluxes that are highly variable in both space and time. Use these systems to test and calibrate models, as well as in field experiments where rainfall, temperature, and other environmental factors are manipulated to understand the interacting effects of environmental change on fluxes.
- Improve approaches to greenhouse gas mitigation by quantifying differences among candidate biofuel management systems with respect to energy as well as nitrous oxide, methane, and other greenhouse gas balances.

2.2. Water Quality, Demand, and Supply

Numerous human activities including industrial processes, urbanization, timber harvest, construction projects, agriculture, and landscaping projects affect water quality (National Research Council 2008). Discharges from these activities contribute to varying degrees of water-quality problems with local and downstream effects on rivers and water bodies. An expansion of biomass and biofuel production will likely affect water quality, demand, and supply. Impacts will vary considerably by region, depending upon competition for water supply; the kind of biomass feedstock and the way it is managed; characteristics of the land; local climate, including potential future changes; and methods used to convert biomass to biofuels.

Some choices of crops and cultivation options could cause soil and nutrient loss and require large amounts of irrigated water. However, options that include cellulosic feedstocks such as woody vegetation (e.g., intensive, short-rotation forestry) and perennial herbaceous species (e.g., switchgrass) have the potential to be produced and harvested in ways that reduce water runoff, soil erosion, and nutrient and pesticide exports to surface and ground waters. Prior research on cellulosic feedstocks has focused on optimizing growth conditions and feedstock productivity. Future research is needed on the impacts of biomass production on water quality and availability. The current lack of knowledge limits our ability to make decisions on the efficient use of water, the control of runoff, and the ability to assess water-quality and water-supply implications for the different cellulosic feedstocks that will be suited to different growing conditions around the country.

Water Quality

Current agricultural practices impact the quality of the nation's water supplies. The extent of sediment and nutrient loss from fields is largely determined by management practices. Practices such as tillage and annual crop production on erodible lands can cause erosion and sediment deposition. Conservation tillage, the integration of perennial cover crops between the rows of annual crops, and the use of native grasses as vegetative filter strips and riparian buffers surrounding annual crops can substantially reduce nutrient and sediment export in agricultural watersheds.

Much can be learned about land-use designs, site preparation, and use of conservation management approaches to reduce surface runoff, erosion, and the export of sediments, nutrients, and pesticides from biofuel feedstock crops (Biomass Research and Development Board 2008). This should be linked to research on crop growth including soil-related processes that enhance plant nutrient availability and reduce input losses.

Watershed-scale models have been used to predict water-quality changes resulting from conversion of corn or other annual crops to switchgrass in the Midwestern United States. (Biomass Research and Development Board 2008; Vadas, Barnett, and Undersander 2008; Nelson, Ascough, and Langemeier 2006). Model results for Iowa, Kansas, and the upper Mississippi River basin suggest that 17 to 43% of current cropland could be converted to switchgrass, reducing erosion by 20 to 90% and decreasing nitrogen and phosphorus export up to 60% if fertilizers are not used. However, models indicate that nitrogen and phosphorus export from switchgrass fields is highly dependent on the amounts of fertilizer applied. When excessive fertilizers are applied to switchgrass fields, nutrient export is comparable to that seen in row crops. Watershed-scale research is needed to assess the aggregated impacts of agricultural production and conservation systems (Richardson, Bucks, and Sadler 2008) and to determine the impacts of incorporating bioenergy production.

Although there is a long history of research on the impacts of forest management on water quality, with the emergence of the biofuel industry, relatively few studies have examined the water-quality relationships of forests managed specifically for bioenergy production. Conversion of unmanaged forests to biofuel production could produce negative effects depending on where these lands are located and how they are managed. An East Texas study of intensive forestry impacts indicated significant increases in storm runoff, erosion, and nutrient loss relative to controls, but the impacts were highly variable over time (harvest cycle, weather) and with different management practices (site preparation, burning) (McBroom et al. 2008a; 2008b).

Water Demand and Supply

U.S. agriculture is the second-largest consumer of water from aquifers and surface supplies. The future biofuel production industry will create new demands on the quantity of water used by agriculture and production forestry. Globally, commercial bioenergy production is projected to consume 18 to 46% of the current agricultural use of water by the year 2050 (Berdes 2002). Population growth and changes in land and how it is used will influence future demands. Water requirements for processing biomass into biofuel also are important, but the quantity of water consumed by processing facilities is considerably less than that consumed by crop cultivation and thus was not a focus of the workshop.

In many parts of the United States, the agricultural sector already faces water shortages. In the arid West, agricultural withdrawals account for 65 to 85% of total water withdrawals. In the East, supplies are under pressure from competing uses, especially in periods of drought. Although overall withdrawals in the United States have decreased since 1980 and efficiency improvements are still possible in irrigation, the amount of water needed for a

Sustainability of Biofuels: Future Research Opportunities 21

biofuel-based energy supply is much greater than equivalent fuel production from fossil fuels.

The understanding needed to assess future impacts of cellulosic feedstock production on the water supply will require investigation of mixed feedstock production systems that vary by location and could be difficult to monitor. Although some water inputs from rainfall or irrigation are incorporated into crop biomass, water is lost primarily through plant transpiration, evaporation, runoff to surface waters, and deep percolation beyond the reach of plant roots. Evapotranspiration rates vary by feedstock, genetics, and weather. Current watershed models may not capture these field-scale effects at the basin scale.

Research is under way at the watershed-scale level to develop the methods needed (Steiner et al. 2008) to understand the implications of future biofuel production on systems and make science-based decisions that will lead to greater sustainability. Also, results of forest conversion experiments from long-term monitoring catchments (e.g., gauged catchments on experimental forests within the U.S. Forest Service) are providing historical data that can be used for improved models. Research is needed to expand methods and information systems to extend evapotranspiration, runoff, and infiltration models from watershed scales to greater regional scales across the entire country. Furthermore, the combination of life cycle analysis and environmental cost accounting with watershed hydrological and water-quality modeling will provide improved tools for analyzing the water requirements of feedstock supplies as well as biofuel conversion plants. A critical research need will be examining how the expansion of biofuels and more intensive agriculture will affect the water cycle and future precipitation patterns, especially within the context of the uncertainty in future climate change.

Research Opportunities

- Understand water-supply requirements to improve prediction and management.
  - Develop hydrological models that reflect the effects of converting agricultural crops, forests, and other land uses to bioenergy feedstock production under a variety of management conditions. Validate model predictions with data obtained from field and watershed studies.
  - Determine the influence of future climate change scenarios on hydrology and bioenergy production. Determine the potential impact of landscape alteration due to fuel crop conversion on local precipitation and other weather variables.

- Understand the impact of biofuel production on water quality to improve prediction and management.
  - Develop field trials that generate near real-time data for identifying the impact of bioenergy crop production on water-quality parameters, and expand hydrological models to include these new data.
  - Link research and modeling on water quantity and quality with information on soil processes and crop growth to more accurately predict the effects of biomass management options.
- Improve approaches to bioenergy feedstock management.
  - Develop new approaches to agricultural and silvicultural land-use design and management practices that reduce runoff of sediments, nutrients, pesticides, or other inputs.
  - Develop integrated decision-making tools at farm, regional, watershed, state, and national levels by integrating data from appropriate spatial and temporal scales of water use, supply, and quality.
  - Determine how site preparation, management, and harvesting strategies for crops and forestlands can be done to minimize erosion and sediment loss.

2.3. Biodiversity and Ecosystem Services

Previous sections discussed ecosystem services such as soil fertility and crop productivity, control of greenhouse gases, and water supply and contamination. Rural landscapes provide many other basic ecosystem services that will continue to be crucial to maintain and improve as the biofuel sector emerges. For instance, a diverse biofuel production system including native species could increase local biodiversity and provide suitable habitat for organisms such as wildlife and predatory or pollinating insects that are beneficial to agricultural and natural ecosystems.

Ecosystem biodiversity often has been associated with a broad range of services, and the idea that more diverse ecosystems sustain greater productivity and system stability is an appealing concept (Shennan 2008). However, the degree to which ecosystem biodiversity is impacted—for better or worse—by the inclusion of biofuel production into existing rural landscapes still needs to be understood to ensure not only sustainable production of biofuels but also food, fiber, and feed.

Sustainability of Biofuels: Future Research Opportunities

Research into the effects of integrating bioenergy production into U.S. agricultural systems provides an opportunity to rethink the structure and function of agricultural landscapes. Sustainable approaches to biofuel production may require diversified and highly integrated management systems to produce the mass of cellulosic feedstocks necessary while at the same time providing needed goods, services, and values from the same working landscapes (Cassman and Liska 2007). Current perennial feedstock management alternatives range from low-diversity systems using a single species (e.g., switchgrass and *Miscanthus*) that produce the greatest biomass per unit area (Schmer et al. 2008) to greater-diversity systems (e.g., mixed forest or grasslands) that may produce lower yields but provide increased ecosystem services (Tilman, Hill, and Lehman 2006; Wallace and Palmer 2007). Research will help understand how these management alternatives compare for a wide range of ecoregions and in conditions in which biofuel feedstocks are likely to be produced. Even in existing agricultural landscapes, few studies have sought to quantify the value of natural landscape components that support ecological services such as wildlife habitat maintenance (McComb, Bilsland, and Steiner 2005). Further research is necessary to value ecosystem services, even though many may not be amenable to monetization or even quantification (Mitchell, Vogel, and Sarath 2008). Quantified measures of ecosystem services will be easy to include in decision models. However, one goal for the investigations should be to generate results in a form amenable to decision making that does not rely on quantifying tradeoffs.

Ultimately, the capacity and feedstock flexibility of cellulosic biofuel refineries may be drivers of changes in landscape structure and therefore significant determinants of biodiversity and ecosystem services. If biorefineries are optimized for a single feedstock, this could tend to reduce landscape diversity and ecosystem services within the feedstock supply area. However, win-win scenarios could be envisioned in which integrated production and processing of multiple cellulosic feedstocks enhance ecosystem services. For example, research could help guide strategies that augment ecosystem services through such approaches as planting small amounts of perennial vegetation grown for cellulosic biomass and strategically located as parts of conservation or riparian buffers that also enhance water quality, pollination, and biocontrol. At larger scales, adding perennial crops could help protect critical habitat corridors.

The impacts on ecosystem services of biomass crops that may become invasive are uncertain. Research is needed to predict the potential impacts as well as to reduce risks.

24 USDA and United States Department of Energy

Research Opportunities
- Identify ecosystem goods, services, and values provided by biofuel feedstocks.
  - Determine ecosystem services for different ecosystems where agricultural and forest feedstocks are likely to be produced. Ensure that the investigations are applied to diverse landscapes, including those dominated by unmanaged landscape components, and areas where food, feed, and fiber are primarily grown.
  - Explore the links among diversification of agricultural landscapes, resilience, and provision of ecosystem services to guide management.
  - Analyze the impacts of climate change scenarios on ecosystem services across a broad range of biophysical and ecological conditions.
- Develop ways to increase ecosystem services.
  - Develop quantitative models and decision tools to evaluate the service, including to monetize it where possible, and bundle ecosystem services to help identify management tradeoffs and synergies, guide more sustainable production decisions, and decrease unintended consequences.
  - Develop harvesting techniques that gather feedstock from timber stands with minimal impact on ecosystem resilience and services.
  - Investigate the potential invasive or gene transfer consequences of introducing new or transgenic bioenergy feedstocks. Develop options to reduce any risks that introduction of these into production systems may present.

2.4. Integrated Landscape Ecology and Feedstock Production Analysis

By its very nature, bioenergy sustainability will require a systems perspective of landscapes across scales—from fields and farms to watersheds and larger regions. Landscape ecology is the study of relationships between spatial patterns and ecological processes for a multitude of scales and organizational levels.

Sustainability of Biofuels: Future Research Opportunities 25

Regional Perspective

To date, bioenergy research has emphasized investigations at small-plot, farm, or field scales and, to some extent, very large scales such as political, national, and global scales. There is a gap, however, in the middle scales, from watersheds to larger regional scales, that may be the most relevant to environmental issues in general and sustainability issues in particular (Robertson et al. 2007). In many cases, current models have not been evaluated for their suitability across varying ecoregions and at different spatial scales of resolution. Furthermore, few field studies have been conducted into the effects of bioenergy feedstock production on watershed quality for ecoregions where feedstocks could be sustainably produced. Without field-based research and the validation of model results, these deficiencies could pose major challenges to the design of biofuel production systems that actually are sustainable. The chief barriers are the lack of knowledge regarding how different processes interact at different scales of resolution, validated model results to interpret impacts across broad ecoregions, and decision tools to direct the development of sustainable management practices and systems. Interdisciplinary research teams involving scientists from the agricultural, forestry, ecological, socioeconomic, and information systems communities will be required to fill such knowledge and technology gaps and provide integrated solutions that effectively target specific components at the appropriate spatial scales. Principles and processes for how human and natural resource systems interact need to be better understood, especially in view of regional landscapes that contain a mosaic of farming and forestry activities, natural areas, and communities.

Model Integration

Improved models and analytical frameworks are needed that integrate biophysical and ecological processes at regional scales, together with economic and other aspects of human behavior. Mechanistic models of crop growth and yield, carbon sequestration and greenhouse gas fluxes, water quality and hydrology, and biodiversity benefits have been developed at plant and field to small regional scales.

Economic models have been developed to capture the impacts of landowner choices with respect to what to grow and how to grow it, including how changes in quality of the natural resources feedstock supply affect prices (Johansson, Peters, and House 2007). Biophysical and economic models have just begun to be combined for analyzing the environmental and economic impacts of technology and policy alternatives and to optimize multiple management objectives (Whittaker et al. 2007). Such integrated analyses are needed to ensure the sustainable use of agricultural landscapes as implementation details and potential tradeoffs will differ across regions. If not fully integrated, these tools used alone may not capture important feedbacks and interactions (Antle and Capalbo 2002). Additionally, the continued development of datasets is necessary to validate model results at regional and other scales (Sadler et al. 2008; Steiner et al. 2008), to ensure the sustainable use of agricultural landscapes as implementation details and potential tradeoffs will differ across regions.

Research Opportunities
- Investigate landscape ecology at regional scales to understand the relationships among diverse processes.
  - Develop analytical frameworks for regional-scale ecological models. Link these models with biophysical and economic models to understand how key aspects of bioenergy production affect the multifunctional roles of agricultural and forest landscapes.
  - Develop regional models that enable the evaluation of management options for climate change scenarios.

3. ECONOMIC DIMENSIONS OF BIOFUEL SUSTAINABILITY

Economic research can help in determining the direct cost competitiveness of cellulosic biofuels with competing sources of transportation fuels. However, the production of biofuels has direct implications for a wide variety of ancillary environmental and social consequences. Expanded economic research will be critical for better understanding the connections among these ancillary impacts and will help support policies leading to development of a biomass sector that creates incentives consistent with society's values.

3.1. Economic and Market Impacts

Several potential sources of cellulosic feedstocks include agricultural crop and forestry residues, perennial grasses, and short-rotation woody crops. An

Sustainability of Biofuels: Future Research Opportunities 27

active area of economic research is to determine the mix of cellulosic feedstocks likely to be competitive in different regions, the spatial pattern of land-use changes these feedstocks will induce, and their implications for crop production and management and for commodity and food prices. Current economic research suggests the possibility of considerable spatial heterogeneity in optimal choice among different feedstock crops, and often a mix of cellulosic feedstocks is likely to be selected. Yields of cellulosic feedstocks are critical determinants of their economic viability (Perrin et al. 2008). Furthermore, to envision a viable biobased energy system, bioenergy crops must compete successfully with traditional food, feed, and fiber crops and with conventional petroleum fuel sources. Farmers will produce cellulosic feedstock crops only if they can receive an economic return at least equivalent to returns from the most profitable alternative crops.

Current economic models of agricultural production can estimate national and regional feedstock production; the role of livestock production; and input demands such as land, fertilizers, and tillage and other production practices. These models also can predict the impacts of increasing biofuel production on commodity and food prices. Although existing models can predict aggregate land-use change, more precise estimates—particularly at smaller scales—will require additional data and research.

A broad assessment of economic impacts requires looking at both the supply side and demand side of markets. Much of current economics research is focused on the supply-side implications of ethanol production (e.g., for agricultural producers, land use, and crop production). A broader assessment also requires consideration of the demand side to determine how consumers' preferences influence outcomes. For example, the demand for biofuels will depend in part on the availability of flex-fuel vehicles, the cost and availability of biofuels and blended fuels, and market prices for gasoline. Economic models need to incorporate the determinants of demand for biofuels under various scenarios of substitutability of biofuels and oil to assess the impact on prices and biofuel use.

Finally, economic analysis of biofuels should consider the implications of biofuel production, prices, and resource use within the global market. For example, the potential for Brazil to substantially expand production of sugarcane ethanol could influence the competitiveness of cellulosic biofuels in the United States. Global economy-wide models examine commodity production at country and subcountry scales and provide insight into direct and indirect land-use impacts of U.S. and international biofuel production. The driving forces behind land-use change in many parts of the world, however,

28 USDA and United States Department of Energy

are linked to political, biophysical, cultural, infrastructure, land-tenure, and social factors in addition to commodity markets. Research will enable modeling of such diverse factors.

External benefits and costs of biofuels, such as environmental consequences, are unlikely to be included adequately (if at all) in private-sector decisions about biofuel consumption and production in the absence of government policy based on sound science. Decision models need components to represent the effects of choices such as feedstock crops, tillage practices, and nutrient applications as well as the impact of agricultural and forest residue removal on soil quality.

Economic analysis can contribute to the design of policies that can address sustain-ability concerns at least cost to society. Moreover, economic analysis also is needed to examine the social costs and benefits of existing biofuel policies such as mandates, tax credits, and import tariffs. For example, the extent to which biofuel mandates reduce gasoline consumption and mitigate climate change depends on a number of parameters that capture human behavior. These include the responsiveness of ethanol and gasoline supply to prices, the extent of substitutability between ethanol and gasoline, and the responsiveness of fuel demand to higher fuel prices.

3.2. Forestry Economics and Land Use

The development of a cellulosic biofuel sector raises a number of questions regarding the future structure of forests and the flow of multiple benefits from these systems. New facilities are just beginning to compete for raw materials in some areas. Although new facilities are being constructed where these materials appear plentiful, not all standing biomass can be considered "available" for timber harvest. Rather, harvest choices and the supply of forest biomass depend on the preferences of private landowners who control the vast majority of commercial timberland in the United States.

Research is needed to ascertain the extent of the potential supply in all forest-producing regions. Research also is needed to address landowner preferences for timber-based revenue versus nontimber amenity values of forests and the implications for aggregate timber supply. In addition, because forest biomass already is used in so many other production processes, understanding the full structure of supply that addresses the complementarities and substitutability of fuel stocks with current production of sawlogs, pulpwood, poles, and other products will be important. This is fundamental to

Sustainability of Biofuels: Future Research Opportunities 29

understanding the potential coevolution of all wood-using sectors. Another important element is the competition of agricultural and forest-based production processes for biofuels and the potential for land-use change in response to changing returns from agricultural and forestry products. Also, both forest and agricultural land are competing with exurban development in many regions, adding another dimension to land-use issues associated with biofuel sustainability.

3.3. Integration of Economic and Biophysical Feedstock Modeling

Research underlying the development of models that fully represent the economics of biofuels requires an interdisciplinary approach. It should integrate biophysical models of feedstock production involving plant, soil, and other ecosystem processes with economic models of production and human behavior. These integrated models should reflect soil carbon biosequestration and take into account the spatially variable nature of agricultural production and environmental quality. Some options, such as selecting perennial crops and placing biorefineries, imply multiyear consequences. New research should represent both economic and environmental effects at regional scales that best capture sustainability issues. Although current economic models can indicate changes in the type and location of production, less is known about the direct impacts of changes on soil and air quality, water use and quality, wildlife habitat and biodiversity, and other environmental considerations.

Such integrated modeling also can play an important role in helping direct biofuel production toward a sustainable future by providing estimates of the social costs and benefits of various policies. Integrated modeling can inform decision makers about the design of alternative policies that consider incentives for reducing greenhouse gas emissions, water-quality degradation, and loss of biodiversity.

Global information on the availability and productivity of land for feedstock production is needed to estimate the potential supply of cellulosic feedstocks in competition with other uses. In particular, identifying the amount of underutilized rural land available for expanding crop production around the world vis-a-vis existing forests, nature reserves, urban areas, and current harvested areas is necessary to determine the implications of indirect land-use changes on greenhouse gas emissions in other countries.

3.4. Life Cycle Analysis

To understand the sustainability of a biofuel, knowing the effects of production and consumption throughout the entire biofuel system is critical. To model the entire system, defining system boundaries is important in including as many relevant factors as possible. Life cycle analysis (LCA) is one of the methods used to conduct these kinds of assessments. This particular approach is especially important to the sustainability of biofuels; it has been addressed explicitly in legislation and will be used by the U.S. Environmental Protection Agency to aid in sustainability assessments. The construction of LCAs for multiple feedstock and conversion technologies of proposed candidates for regionally significant production should include well-to-wheel approximations of net energy production (Schmer et al. 2008). Other considerations needed are mass balance analysis of irrigation water and precipitation, land use, nutrients, and agrichemicals associated with prospective feedstocks and conversion technologies.

As traditionally defined and practiced, using LCA to capture some of a system's critical complexity is difficult. Substitutions among existing technologies in response to relative price changes and technological changes that increase efficiency are not captured by current LCA. It also does not analyze indirect land use, is not dynamic, and does not easily accommodate multiple changes simultaneously.

However, performing LCA is important for the evaluation of biofuel sustainability. These analyses should be conducted at a variety of geographic scales. For instance, additional research is needed to determine the net energy requirements of larger-scale feedstock cultivation and biofuel processing facilities to better define the efficiencies possible in cellulosic biofuel production systems. Assumptions involved in LCA should be transparent so that fair comparisons can be made across different biofuel technologies. For example, some biofuel technologies may be more carbon saving while others may reduce oil imports. Direct and transparent analyses enable an assessment of these kinds of tradeoffs. Sensitivity analyses should be conducted to explore alternate assumptions and parameters, with uncertainties in underlying processes and parameters clearly set forth. LCA models should accommodate risk analysis and flexible representations of various input parameters. Furthermore, a variety of potential users would benefit from user-friendly models.

Comparisons of the greenhouse gas results from LCAs can be difficult because biofuel production systems tend to be complex, and the scope,

Sustainability of Biofuels: Future Research Opportunities   31

parameter values, methodologies, and assumptions about energy inputs or credits (e.g., the potential for electricity cogeneration) and other factors tend to be uniquely defined for each study (Liska and Cassman 2008; Dale et al. 2008). To make LCA useful for these comparisons, approaches should be standardized (Wallace and Mitchell 2009). LCA modelers should determine how to set standards and practices such as similar treatment of a common set of variables.

3.5. Emissions from Land-Use Change

A major limitation to current LCA models is the inability to measure and account for greenhouse gas emissions from land-use change. Methods are needed to rapidly measure greenhouse gas emissions across variable landscapes so that the effects of land-use redirection to biofuel production can be determined. These should include emissions from the clearing of forests, grasslands, and other natural ecosystems to produce biofuels or other agricultural crops displaced by biofuels (e.g., Fargione et al. 2008). LCA models should be developed that assume realistic soil management practices and accepted technologies. These technologies include the use of conservation tillage or low-disturbance systems and the retention of proper amounts of crop residue on the soil surface so that impacts on existing levels of soil organic carbon are minimized even after several years of cropping (Follett et al. 2009).

Realistically assessing the impacts of biofuel production in the United States based on land-use choices elsewhere is especially difficult because potential land-use change is influenced by many different factors, including the expansion of roads and infrastructure into undeveloped lands for other purposes, changes in the values of agricultural and wood products, developing technologies, and sovereign choices. Drivers of land-use change are not single actions but complex interactions among cultural, economic, technological, political, and biophysical forces (Dale et al. 2008).

Methodologies for quantifying changes in land use that are attributable to biofuel production are still in the early stages of development.

3.6. Economic Risks and Uncertainty

The uncertainty of future events such as severe weather, climate change, or dramatic changes in oil prices can have important consequences for decisions spanning long periods of time. Assessing the level, source, type, and location of risk associated with various uncertainties is an important consideration in economic research. At the level of farm decisions, annual crops provide some flexibility not present in perennial biofuel crops. Decisions by farmers are sensitive to expectations regarding biorefinery investments and operations, future improvements in crop varieties, petroleum prices, and many other multiyear changes. Production of cellulosic feedstocks imposes new risks on producers and refiners because it can involve decision making and contractual commitments over many years. Uncertain market prices for energy crops and lack of other market outlets for those crops can make energy-crop profits dependent on uncertain or volatile oil prices and on the location of biorefineries.

The uncertainty caused by possible rapid innovations leading to new, genetically superior varieties of energy crops or improvements in conversion technologies also could influence investment decisions.

These multiyear decisions are common in economic analysis, for example, option theory and decision theory. Accommodating risk and uncertainty in much of the current biofuel analysis, however, will require research. Properly representing risk will be particularly important if decisions are to be made regarding potential governmental policies. For sound public policy designed to support a sustainable biofuel industry, research is needed on the implications of alternative models of contracting for feedstock, providing crop insurance, and other risk-mitigating and government-based safety nets.

New biofuel crops may provide new sources of revenue and enhanced job prospects for selected rural areas. Biorefining may provide further economic opportunities and stimulate growth. Economic opportunities for farm and rural communities need to be considered in the context of farm and off-farm economics and finance. The extent to which rural communities will capture these economic benefits is still unknown. Economic modeling would help predict, for example, how different parties might capture the profit and bear the risk in the biofuel value chain and estimate the multiplier effect that forecasts job growth.

Research Opportunities

Economic modeling applied to sustainable biofuel production is still nascent. It is driven by the need to supply information to a variety of decisions as well as to serve as input for analysis by other experts.

Sustainability of Biofuels: Future Research Opportunities 33

Some needs include the following:

- Estimate the quantity and cost of biofuel production.
  - Develop regional and aggregated biofuel supply models that integrate the diversity of energy feedstocks, growing conditions, ecosystem services, and economic parameters.
  - Develop scenarios of patterns of biomass crop selection and cultivation at national, regional, and local levels. Provide these scenarios to analysts and decision makers in related policy areas such as climate change, international trade, water quality and demand, and rural development.
  - Model and analyze the economic implications of biofuel production, prices, and resource use within the global market. Consider international markets and trade in the production of biofuels.
  - Expand supply model capability to include analysis of climate change scenarios that could affect long-term growing conditions.
- Evaluate the amount of noncropland available in the United States and other parts of the world.
  - Assess and quantify competing land use and examine forces that cause indirect land-use effects.
  - Investigate factors (economic and noneconomic) that influence the potential conversion of different land uses or covers to feedstock production.
  - Use information from ecological studies and models to assess the availability and value of land, water, and other natural resources.
- Analyze ancillary benefits and disbenefits of biofuel production through life cycle analysis.
  - Expand life cycle analysis to capture critical processes and parameters that will enable the examination of well-to-wheel biofuel production. Provide ways to develop standard assumptions so comparisons of different LCAs can be more transparent and accomplished easier.
  - Quantify the factors responsible for land-use change to assess the carbon fluxes appropriately attributable to the establishment of biofuel cropping systems elsewhere. Incorporate indirect land-use effects as appropriate into LCA models.
- Develop decision tools to enable encouragement of the adoption of onfarm practices to meet environmental objectives.

34 USDA and United States Department of Energy

- Analyze the economic effects of policy options that provide economic incentive mechanisms and encourage compliance with environmental objectives.
- Provide information on future infrastructure requirements.
- Analyze demand for flex-fuel vehicles, filling stations, and ethanol transportation.

4. SOCIAL DIMENSIONS OF BIOFUEL SUSTAINABILITY

Previous sections of this report discussed environmental and economic dimensions of cellulosic biofuels and related research needs. This section outlines the science agenda required to understand the social and technological changes needed to achieve sustainable biofuels, including their implications for farmers, foresters, rural communities, and other stakeholders. The biofuel sector cannot develop sustainably without an understanding of and effective response to stakeholder values, choices, behaviors, and reactions, along with careful consideration of the social structures and policies influencing that development. Research and engagement in this arena will provide policy and decision makers with information needed to support decisions at individual, community, and national scales.

4.1. Understanding Stakeholder Needs and Motivations

Everyone has a stake in national energy security and in sustainable biofuel development, but farmers and foresters; rural community decision makers; the biofuel industries; and local, regional, and national policy makers will play pivotal roles in achieving a sustainable future. A wide range of motivations drives stakeholder values, choices, and behaviors: price signals, resource and equipment needs, infrastructure requirements, environmental protections, number and quality of jobs, lifestyle changes, economic multipliers, and policy incentives. Understanding how stakeholders view, evaluate, and make choices about potential opportunities and risks of biomass and biofuel development is essential to designing and managing systems that capitalize on the opportunities while avoiding or mitigating unintended adverse consequences.

Sustainable feedstock production and biofuel development have the potential to fundamentally alter the management choices and practices of farmers and foresters, changing agricultural and forestry landscapes as well as rural communities. Growing feedstocks for biofuels presents new job opportunities in biomass production, transport and storage, biofuel processing, and ancillary services and industries, but these changes are likely to place increased demands on essential resources and systems needed for food production, power supplies, transportation, and water quality and quantity. The extent of change in rural demographics and development, as well as the effects on farmers, foresters, rural communities, and other stakeholders, will depend on many things, including land tenure, individual and regional land-use decisions, workforce development, community capacity, biorefinery ownership, biomass processing choices, and whole systems designs.

4.2. Building on Lessons Learned from Biofuel Production at Home and Abroad

The nation has nearly two decades of experience with grain-based biofuels. Corn ethanol and soy diesel, which have contributed significantly as substitutes and additives for fossil fuels, have brought us to our current threshold of cellulosics. Yet, intensification of grain production for biofuels has raised a number of concerns relevant to long-term sustainability including resource competition, soil and water degradation, wildlife habitat and conservation disturbance, gulf hypoxia, disruption in the livestock industries, and escalating food prices. Learning from the U.S. experience with grain-based biofuels like corn ethanol and soy diesel will be important to avoiding past mistakes that could threaten the successful expansion of cellulosic biomass and biofuel production.

Only through a nuanced understanding of the social dynamics surrounding biofuel production will producers, policy makers, and community decision makers be able to design systems that prevent or mitigate unintended adverse consequences. This will include understanding the relative social merits and shortcomings of various biomass alternatives at different spatial and temporal scales, understanding the social advantages and disadvantages of biomass monocultures and polycultures and deliberating the tradeoffs, and understanding the relative social benefits and risks of diversified biofuel production enterprises. Inherent in all this is evaluating or examining how the many stakeholders identify, value, and weigh the social costs and benefits, negotiate the myriad tradeoffs that will be required, and respond to the consequences.

Just as the lessons learned in the U.S. experience with grain-based biofuels are crucial to future planning and development, so too are the lessons learned by other biofuel efforts in Europe, South America, and elsewhere. Insights from international experiences will provide a broader view of numerous issues including interconnections among global resource inventories, competition, and responses; potential effects of different policy and incentive options; stakeholder acceptance of new technologies; environmental risks and protections; and development of effective practices in agriculture, forestry, community planning, and facility siting.

4.3. Understanding the Social Effects of Scale and Complexity for Biofuel System Design

Farms, woodlots, forests, and communities vary markedly by size, complexity, climate, geography, resource endowments, and human capital. Therefore, research and design of cellulosic biomass production need to reflect this diversity as well as the values and capacities of rural communities, biofuel and agricultural industries, and other stakeholders. Analysis and understanding must encompass multiple scales and complexities—from individual farms and forests to whole communities and ecosystems—so science can inform decision making and design at local, regional, national, and global levels.

The effects of scale are not limited to cellulosic biomass production but extend also to biofuel development and systems design. Scale and ownership patterns will influence the design and siting of biofuel facilities. The scale of facilities will affect and be affected by industry concentration, community capacity, infrastructure support, transportation and storage costs, workforce potential, and income generation. The environmental outcomes that differ by feedstock and processing facility scale and complexity will influence public perceptions and support. Research is needed to identify biorefining systems and enterprise structures that optimize benefits and reduce undesirable consequences for biomass producers and communities and society at large. The implications of ownership and scale for cellulosic feedstock production, conversion technologies, and biorefining systems require examination, as do the development options they permit.

4.4. Understanding Social Dynamics, Human Choices, Risk Management, and Incentives

The pursuit of sustainable biofuel production will entail many decisions, negotiations, compromises, and tradeoffs on the part of feedstock producers, rural communities, biofuel industries, and society generally. Agricultural and forestry biomass providers will make feedstock choices constrained by various geographic and climatic conditions, resource availability, equipment requirements, establishment costs in time and investment, labor demands, and their own technical and financial capabilities. Land tenure constraints, risk management options, and enterprise goals also will influence choices and management practices of feedstock producers. Their decisions and behaviors will depend on whether they own or rent land and, if they rent, whether their landlord is a family member, a neighbor, a rural community member, an urban absentee owner, or a corporation. Other factors influencing producer decisions are whether they have authority to make short- and long-range decisions on land use, resource management, cropping choices, and equipment investments; their level of indebtedness; the risk management tools at their disposal; and the types and duration of contracts, subsidies, and conservation programs to which they have committed. Likewise, their decisions and practices will be shaped by the goals of their farming and forestry enterprises. A goal to support a family through farming or to bequeath land to a new generation of farmers will influence decisions and behaviors differently from a goal to sell the land, resources, or enterprise to developers. Choices made on farm or forest enterprises can be driven more by short-term profitability than long-term viability, or by some balance of both. Decisions of biomass producers will depend on the unique conditions defining each scenario.

The decisions and decision-making processes of communities are similarly complex. Rural community members, as individuals and in aggregate, evaluate options and make development decisions based on available resources and their capacity and willingness to support, finance, and invest in new futures. They determine their level of infrastructure needs and support and make investments through zoning, taxes, incentives, and policy in an effort to attract business; support production; generate revenue, jobs, and income; and maintain viability. Workforce development needs and capacities are central to their planning as they assess current capacities, examine prospects for growth or development, and mobilize their efforts. Resource inventories provide a window on the extent of resource competition against which alternative uses must be weighed and difficult options negotiated. The

38 USDA and United States Department of Energy

development and quality-of-life goals of communities and their constituent members will differentially shape the decisions and actions they make to pursue cellulosic feedstock and biofuel development.

Feedstock providers and rural communities are not the only stakeholders making decisions in this arena, of course. Biofuel industries are actively engaged in feed-stock inventories and assessments of resource availability. They evaluate infrastructure needs and availability and consider flexible conversion technologies and facility siting accordingly. Their decisions are constrained by the sufficiency of these resources, as well as by financial, economic, and political considerations. Their ability to leverage resources, negotiate contracts, and influence community decision makers, as well as their capacity to handle logistics and operate processing competitively, all shape the kinds of decisions and investments that industry makes.

Sustainable biofuel development is motivated by at least five larger societal goals: improved farm and forest economies and producer well-being; rural economic development; global climate change mitigation; energy security; and national security. At times these goals may not seem complementary or compatible. Understanding whether and how different stakeholders can reach consensus about these goals and the paths to achieve them is an essential focus of the social sciences and the decision and risk management sciences, as well as the planning and design disciplines. It will be important to identify societal outcomes at a variety of scales and for various scenarios, understand economic and political constraints, assess the commitment of different sectors of society, and sort out the values, compromises, negotiations, and tradeoffs that will be necessary to attain these outcomes. Research and understanding of these social processes can provide policy and decision makers at all levels with information essential to policy development and negotiation, incentive creation and evaluation, and design of support mechanisms and structures to best serve larger societal goals.

Finally, information access is another critical issue. Stakeholders of all kinds—be they feedstock providers, community planners, industry professionals, or national policy makers—will need to find information supported by environmental, economic, and social science research to identify the many alternatives, evaluate options, weigh likely outcomes, and make decisions. A diverse portfolio of research, decision aids, education, communication tools, and outreach and extension activities will be needed to inform and support their decisions. These are among the many social science challenges that will require future investment by the research and development communities committed to achieving sustainable biofuel futures.

Sustainability of Biofuels: Future Research Opportunities

Research, Education, and Extension Opportunities

Addressing the challenges underlying the human and social dimensions of cellulosic biomass and biofuel development must be a part of any research and outreach program focused on sustainability. The issues raised in this section are concisely captured in the following list of investment opportunities. Although these opportunities are framed from a social science perspective, their solution will require the collaborative efforts of ecologists, biological and physical scientists, systems design engineers, and economists alongside sociologists, geographers, agricultural historians, family and consumer scientists, agricultural educators and communicators, and extension professionals.

Understanding Stakeholder Needs and Motivations

- Identify incentives and impediments to individual farmer and forest landowner decisions to grow energy and alternative crops.
- Identify and understand land tenure characteristics, crop and product attributes, capital and equipment needs, management practices, labor demands, and technical capacities that influence adoption of cellulosic biomass crops by farmers and foresters.
- Identify and understand market characteristics and financial and community support programs that influence adoption of cellulosic biomass crops by farmers and foresters and biofuel facilities by communities and industry.
- Develop specialized risk management tools to assist feedstock providers and biorefining facilities. Identify barriers to new investment and development and commercialization of new products.
- Assess rural infrastructure and workforce development needs and opportunities.
- Analyze the social processes, structures, and institutional arrangements underlying community capacity, vulnerability, and resiliency.
- Investigate and design strategies to create, increase, and retain value from cellulosic biomass and biofuel development for agricultural producers, private forest landowners, and rural communities.

40 USDA and United States Department of Energy

Building on Lessons Learned from Biofuel Production at Home and Abroad

- Conduct comparative, historical, and international research to evaluate the experience with U.S. production of grain-based biofuels and other countries' bioenergy production.

Understanding the Social Effects of Scale and Complexity for Biofuel System Design

- Develop tools, metrics, and design criteria to assess social, economic, and environmental sustainability of cellulosic feedstock and biofuel development at local, regional, national, and global scales.
- Conduct spatial and temporal analyses of farmer and forester adoption, community support and outcomes, industry experience, and societal responses to cellulosic feedstock and biofuel development.
- Develop scale-neutral and scale-sensitive research and technologies to compare differential consequences of variously scaled cellulosic biomass and biofuel production systems and assess the outcome of large industrial-scale, regionally based, and local decentralized energy systems.

Understanding Social Dynamics, Human Choices, Risk Management, and Incentives

- Analyze stakeholder values, evaluations, actions, and responses regarding alternative biofuel crop options, different ownership models, infrastructure requirements, employment options, and environmental concerns to provide the science foundation to inform decision making at all levels.
- Evaluate a range of rural community effects including changes in water use and demand, management of biorefinery wastes, impacts on property value, reactions to increased ownership concentration, fewer farms, and less control of biofuel production systems to provide the science foundation to inform local decision making as well as local, regional, and federal policy making.
- Conduct modeling and analysis to evaluate the efficacy of existing and proposed incentives for cellulosic biofuel development; examine their effects on rural communities; and predict impacts of these options on biomass production, biorefining capacity, bioenergy industry structure, trade, and international markets.

Sustainability of Biofuels: Future Research Opportunities — 41

- Assess potential approaches for developing human capital in anticipation of changing labor requirements.
- Identify and evaluate incentives, investments, and formal or informal educational needs required to nurture biomass producers.
- Develop appropriate education and outreach programs that inform the next generation of cellulosic feedstock producers, as well as biofuel workers and consumers, and provide information to communities for sustainable biofuel development.

5. SUMMARY

The joint USDA-DOE Sustainability of Biofuels Workshop, held October 28–29, 2008, stimulated an interactive discussion among a wide range of experts on the state of the science and research needed to establish sustainable production and utilization of cellulosic biofuels. This report summarizes that discussion and presents a series of new and critically important areas of research. Interdisciplinary teams involving scientists from the agricultural, ecological, socioeconomic, and information system communities will be required to fill knowledge and technology gaps and provide integrated solutions that effectively target specific challenges. This research, however, must maintain a holistic view of the entire biofuel production
system and its socioecological impacts. DOE, USDA, and other federal agencies now have a unique opportunity to use these recommendations to develop an integrated research agenda that addresses the environmental, economic, and social dimensions of cellulosic biofuels across multiple scales and ensures that this emerging industry grows sustainably.

APPENDIX A. REFERENCES

Antle, J. M., and S. M. Capalbo. 2002. "Agriculture as a Managed Ecosystem: Policy Implications," *Journal of Agricultural and Resource Economics* 27, 1–15.

Archer, D. W., and A. D. Halvorson. 2009. "Greenhouse Gas Mitigation Economics for Irrigated Cropping Systems in Northeastern Colorado," *Soil Science Society of America Journal*, in review.

Baldani, J. I., and V. L. D. Baldani. 2005. "History on the Biological Nitrogen Fixation Research in Graminaceous Plants: Special Emphasis on the Brazilian Experience," *Anais da Academia Brasileira de Ci* 77, 549–79.

Berdes, G. 2002. "Bioenergy and Water: The Implications of Large-Scale Bioenergy Production for Water Use and Supply," *Global Environmental Change* 12, 253–71.

Biomass Research and Development Board. 2008. *Increasing Feedstock Production for Biofuels: Economic Drivers, Environmental Implications, and the Role of Research* (http://brdisolutions.com/Site Docs/Increasing Feedstock_revised.pdf).

Boddey, R. M., et al. 2003. "Endophytic Nitrogen Fixation in Sugarcane: Present Knowledge and Future Applications," *Plant and Soil* 252, 139–49.

Cassman, K. G., and A. J. Liska. 2007. "Food and Fuel for All: Realistic or Foolish?" *Biofuels, Bioproducts, and Biorefining* 1, 18–23.

Dale, V. H., et al. 2008. "Interactions Between Bioenergy Feedstock Choices and Landscape Dynamics and Land Use," *Ecological Applications*, in review.

de Resende, A. S., et al. 2006. "Long-Term Effects of Pre-Harvest Burning and Nitrogen and Vinasse Applications on Yield of Sugarcane and Soil Carbon and Nitrogen Stocks on a Plantation in Pernambuco, N.E. Brazil," *Plant and Soil* 281, 339–51.

Fargione, J., et al. 2008. "Land Clearing and the Biofuel Carbon Debt," *Science* 319, 1235–38.

Follett, R. F., et al. 2009. "Conservation-Tillage Corn after Bromegrass: Effect on Soil Carbon and Soil Aggregates," *Agronomy Journal* 101(2), 261–68.

Johansson, R., M. Peters, and R. House. 2007. *Regional Environment and Agriculture Programming Model (REAP)*. USDA-ERS, Technical Bulletin No. TB-1916, 118 pp., March (http://ers.usda.gov/publications/tb1916/).

Jordan, N., et al. 2007. "Sustainable Development of Agricultural Bio-Economy," *Science* 316, 1570–71.

Liska, A. J., and K. G. Cassman. 2008. "Towards Standardization of Life-Cycle Metrics for Biofuels: Greenhouse Gas Emissions Mitigation and Net Energy Yield," *Journal of Biobased Materials and Bioenergy* 2, 187–203.

Lubowski, R. N., et al. 2006. *Major Uses of Land in the United States, 2002*. Economic Information Bulletin No. EIB-14, U.S. Department of Agriculture, Economic Research Service (http://ers.usda.gov/Publications/EIB14).

Sustainability of Biofuels: Future Research Opportunities   43

McBroom, M. W., et al. 2008a. "Water Quality Effects of Clearcut Harvesting and Forest Fertilization with Best Management Practices," *Journal of Environmental Quality* 37, 114–24.

McBroom, M. W., et al. 2008b. "Storm Runoff and Sediment Losses from Forest Clearcutting and Stand Re-Establishment with Best Management Practices in East Texas, USA," *Hydrological Processes* 22, 1509–22.

McComb, B. C., D. Bilsland, and J. J. Steiner. 2005. "Association of Winter Birds with Riparian Condition in the Lower Calapooia Watershed, Oregon," *Northwest Science* 79, 164–71.

Mitchell, R. B., K. P. Vogel, and G. Sarath. 2008. "Managing and Enhancing Switchgrass as a Bioenergy Feedstock," *Biofuels, Bioproducts, and Biorefining* 2, 530–39.

National Research Council. 2008. *Mississippi River Water Quality and the Clean Water Act: Progress, Challenges, and Opportunities*, National Academies Press, Washington, D.C.

Nelson, R. G., J. C. Ascough II, and M. R. Langemeier. 2006. "Environmental and Economic Analysis of Switchgrass Production for Water Quality Improvement in Northeast Kansas," *Journal of Environmental Management* 79, 336–47.

Perrin, R. K., et al. 2008. "Farm-Scale Production Cost of Switchgrass for Biomass," *BioEnergy Research* 1, 91–97.

Polidoro, J. C. 2001. *O Molibdênio na Nutrição Nitrogenada e na Fixação Biológica de Nitrogênio Atmosférica Associada a Cultura da Cana-de-açúcar*, D. Sc. thesis, Universidade Federal Rural do Rio de Janeiro, RJ, Brasil.

Richardson, C. W., D. A. Bucks, and E. J. Sadler. 2008. "The Conservation Effects Assessment Project Benchmark Watersheds: Synthesis of Preliminary Findings," *Journal of Soil and Water Conservation* 63, 590–604.

Robertson, G. P., et al. 2008. "Sustainable Biofuels Redux," *Science* 322, 49–50.

Robertson, G. P., et al. 2007. "New Approaches to Environmental Management Research at Landscape and Watershed Scales," pp. 27–50 in M. Schnepf and C. Cox, eds., *Managing Agricultural Landscapes for Environmental Quality*, Soil and Water Conservation Society, Ankeny, Iowa.

Sadler, E. J., et al. 2008. "Sustaining the Earth's Watersheds—Agricultural Research Data System: Data Development, User Interaction, and

Operations Management," *Journal of Soil and Water Conservation* 63, 577–89.

Schmer, M. R., et al. 2008. "Net Energy of Cellulosic Ethanol from Switchgrass," *Proceedings of the National Academy of Sciences* 105, 464–469.

Shennan, C. 2008. "Biotic Interactions, Ecological Knowledge, and Agriculture," *Philosophical Transactions of the Royal Society B* 363, 717–39.

Steiner, J. L., et al. 2008. "Sustaining the Earth's Watersheds: Overview of Development and Challenges," *Journal of Soil and Water Conservation* 63, 5569–76.

Tilman, D., J. Hill, and C. Lehman. 2006. "Carbon-Negative Biofuels from Low-Input High-Diversity Grassland Biomass," *Science* 314, 1598–1600.

USDA ERS. 2008. *USDA Agricultural Projections to 2017*. Office of the Chief Economist, World Agricultural Outlook Board, U.S. Department of Agriculture Economic Research Service. Prepared by the Interagency Agricultural Projections Committee. Long-term Projections Report OCE-2008-1 (http://ers.usda.gov/Publications/OCE081).

USDA/NASS. 2008. *Crop Production Historical Track Records*, April 2008. U.S. Department of Agriculture, National Agricultural Statistics Service.

Vadas, P. A., K. H. Barnett, and D. J. Undersander. 2008. "Economics and Energy of Ethanol Production from Alfalfa, Corn, and Switchgrass in the Upper Midwest, USA," *BioEnergy Research* 1, 44–55.

Wallace, L. L., and R. B. Mitchell. 2009. "Are Rangeland Biofuel Feedstocks Ecologically Sustainable?" *Ecological Applications*, submitted.

Wallace, L., and M. W. Palmer. 2007. "LIHD Biofuels: Toward a Sustainable Future," *Frontiers in Ecology and the Environment* 5(3), 115.

Whittaker, G. W., et al. 2007. "A Hybrid Genetic Algorithm for Multiobjective Problems with Activity Analysis-Based Local Search," *European Journal Operations Research* 193, 195–203.

Wilhelm, W. W., et al. 2007. "Corn Stover to Sustain Soil Organic Carbon Further Constrains Biomass Supply," *Agronomy Journal* 99, 1–11.

APPENDIX B. LIST OF PARTICIPANTS

Sustainability of Biofuels Workshop
State of the Science and Future Directions
A Workshop Jointly Sponsored by the U.S. Department of Energy and the
U.S. Department of Agriculture; October 28–29, 2008
Bethesda Marriott in Bethesda, Maryland

Opening Plenary Session Presentations by
Senior Federal Agency Officials
Gale Buchanan
Under Secretary for Research, Education, and Economics
U.S. Department of Agriculture
Raymond Orbach
Under Secretary for Science U.S. Department of Energy

Anna Palmisano
Associate Director, Office of Biological
and Environmental Research
U.S. Department of Energy

Steve Shafer
Agricultural Research Service
Deputy Associate Administrator,
Natural Resources and Sustainable Agricultural Systems
U.S. Department of Agriculture

Invited Workshop Participants
Rob Anex
Iowa State University

David Archer
Agricultural Research Service
U.S. Department of Agriculture

Jeff Arnold
Agricultural Research Service
U.S. Department of Agriculture

Randall J.F. Bruins
U.S. Environmental Protection Agency

Juli Brussell
University of New Hampshire

Ken Cassman
University of Nebraska-Lincoln

Alice Chen
U.S. Environmental Protection Agency

John Cromartie
Economic Research Service
U.S. Department of Agriculture

Anthony Crooks
U.S. Department of Agriculture

Bruce Dale
Michigan State University

Virginia Dale
Oak Ridge National Laboratory

Mark David
University of Illinois at Urbana-Champaign

Eric Dohlman
Economic Research Service U.S. Department of Agriculture

Jae Edmonds
University of Maryland

Michael Farmer
Texas Tech University

Ron Follett

Agricultural Research Service U.S. Department of Agriculture

Chelcy Ford
U.S. Forest Service
U.S. Department of Agriculture

Randall Fortenberry
University of Wisconsin-Madison

Thomas Gray
U.S. Department of Agriculture

Nathanael Greene
Natural Resources Defense Council, Inc.

Brenda Haendler
Booz Allen Hamilton

Kathy Halvorsen
Michigan Technological University

Jerry Hatfield*
Agricultural Research Service
U.S. Department of Agriculture

Jason Hill
University of Minnesota

Kyle Hoagland
University of Nebraska-Lincoln

Andy Isserman
Economic Research Service U.S. Department of Agriculture

Madhu Khanna*
University of Illinois at Urbana-Champaign

Russell Kreis, Jr.
U.S. Environmental Protection Agency

Doug Landis*
Michigan State University

Ephraim Leibtag
Economic Research Service
U.S. Department of Agriculture

Richard Lowrance*
Agricultural Research Service
U.S. Department of Agriculture

Al Lucier
National Council on Air and
Stream Improvement, Inc.

Lee Lynd*
Dartmouth College

Jerry Melillo*
Marine Biological Laboratory

R. Michael Miller
Argonne National Laboratory

John Miranowski
Iowa State University

Rob Mitchell
Agricultural Research Service
U.S. Department of Agriculture

Pat Mulholland*
Oak Ridge National Laboratory

Peter Nowak**
University of Wisconsin-Madison

William (Bill) Orts
Agricultural Research Service
U.S. Department of Agriculture

Deborah Page-Dumroese*
U.S. Forest Service
U.S. Department of Agriculture

Bob Perlack
Oak Ridge National Laboratory

Fran Pierce
Washington State University
Heather Reynolds
Indiana University

Phil Robertson*
Michigan State University

John Sadler
Agricultural Research Service
U.S. Department of Agriculture University of Missouri

Theresa Selfa*
Kansas State University

Raghavan Srinivasan
Texas A&M University

Roya Stanley*
Iowa Office of Energy Independence
Bryce Stokes
U.S. Forest Service
U.S. Department of Agriculture

Scott Swinton
Michigan State University

Larry Teeter
Auburn University

Vincent Tidwell
Sandia National Laboratories

Jim Tiedje
Michigan State University

Ron Trostle
Economic Research Service
U.S. Department of Agriculture

Jon Tyndall
Iowa State University

Larry Walker
Cornell University

Linda Wallace
University of Oklahoma

*Participant speaker
**Participant speaker unable to attend; presentation given by a colleague

In: Biofuel Sustainability: Research Areas... ISBN: 978-1-62100-320-5
Editors: L. S. Graver, M. R. Kriss © 2012 Nova Science Publishers, Inc.

Chapter 2

MEASURING THE INDIRECT LAND-USE CHANGE ASSOCIATED WITH INCREASED BIOFUEL FEEDSTOCK PRODUCTION[*]

United States Department of Agriculture

Report to Congress

This repost was prepared by the USDA Economic Research Service and
Office of the Chief Economist, with contributions from:
Elizabeth Marshall
Margriet Caswell
Scott Malcolm
Mesbah Motamed
Jim Hrubovcak
Carol Jones
Cynthia Nickerson

---

[*] This is an edited, reformatted and augmented version of the United States Department of
Agriculture publication, Economic Research Service, dated on February 2011.

ABSTRACT

The House Report 111-181 accompanying H.R. 2997, the 2010 Agriculture, Rural Development, Food and Drug Administration, and Related Agencies Appropriations Bill, requested the USDA's Economic Research Service (ERS) in conjunction with the Office of the Chief Economist, to conduct a study of land-use changes for renewable fuels and feedstocks used to produce them. This report summarizes the current state of knowledge of the drivers of land-use change and describes the analytic methods used to estimate the impact of biofuel feedstock production on land use. The models used to assess policy impacts have incorporated some of the major uncertainties inherent in making projections of future conditions, but some uncertainties will continue to exist. The larger the impact of domestic biofuels feedstock production on commodity prices and the availability of exports, the larger the international land-use effects are likely to be. The amount of pressure placed on land internationally will depend in part on how much of the land needed for biofuel production is met through an expansion of agricultural land in the United States. If crop yield per acre increases through more intensive management or new crop varieties, then less land is needed to grow a particular amount of that crop.

ACKNOWLEDGMENTS

We received helpful comments and advice from Harry Baumes, Jim Duffield, and Hosein Shapouri from the USDA Office of Energy Policy and New Uses; Sally Thompson, Mary Bohman, Keith Fuglie, and Erik Dohlman from the USDA Economic Research Service; and anonymous reviewers from the U.S. Environmental Protection Agency, Office of Management and Budget, Department of Energy, Department of Commerce, and Council Environmental Quality.

EXECUTIVE SUMMARY

In recent years, concerns have been raised about potential domestic and international land-use changes that might be associated with scaling up biofuel feedstock production. Increased competition for productive land and the resulting shifts in land use to produce food, feed, fiber, and fuel have potential impacts on greenhouse gas emissions, biodiversity, and water quality. The

Measuring the Indirect Land-Use Change Associated ... 53

House Report 111-181 accompanying H.R. 2997, the 2010 Agriculture, Rural Development, Food and Drug Administration, and Related Agencies Appropriations Bill, requested the USDA's Economic Research Service (ERS) in conjunction with the Office of the Chief Economist, to conduct a study of land-use changes for renewable fuels and feedstocks used to produce them. This report is a response to that request, and it summarizes the current state of knowledge of the drivers of land-use change and the role of biofuel production in affecting land-use change. The analytical frameworks that have been used to address land-use impacts of increased biofuels production are presented, and the estimation task is explored in more detail. The objective was to survey the literature in a neutral, objective way. There was no intention to suggest that USDA does or does not agree with any assumptions or model results.

In the United States, the most advanced efforts to integrate consideration of land-use related impacts into policy have been associated with the Energy Independence and Security Act (EISA) of 2007 and policies such as the California Low Carbon Fuel Standard (LCFS). The analyses used for these policies require making projections about future values of domestic and international crop supply and demand, population growth, economic conditions, and land-use policies. Each element is a driver of land-use change, and the future value for them is uncertain. Uncertainty is an unavoidable aspect of policy-impact modeling that makes projections into the future. This report explores how some of the major uncertainties have been incorporated within models of land-use change, and how the estimates have changed over time. The publishing of research results has stimulated discussion of basic assumptions, parameters, and increased model transparency.

Continued research and modeling efforts will narrow the bands of uncertainty associated with projections of land-use change and domestic policy. New models, model refinements, and improved data will all help increase the precision with which input parameters are estimated and behavioral relationships are represented. Still, the successful integration of science and policy must come with the recognition that future projections will always carry some degree of uncertainty suggesting that it will need to be accommodated in policy design.

Highlights of the study:

- The larger the impact of domestic biofuels feedstock production on commodity prices and the availability of exports, the larger the international land-use repercussions are likely to be. Given the size of U.S. agriculture's influence on international markets, producers and

54 United States Department of Agriculture

consumers overseas regularly react to market or policy-related events in the United States. Price signals are the critical link between the behavior of the domestic agricultural sector and the induced land-use response of other countries.

- The amount of pressure placed on land internationally will depend in part on how much of the land needed for biofuel production is met through an expansion of agricultural land in the United States. The allocation of current cropland among different crops is based on expected returns in the next growing season. Land conversion between broad land-use categories can be both costly and irreversible and is therefore driven by longer-term economic factors.
- If crop yield per acre increases through more intensive management or new crop varieties, then less land is needed to grow a particular amount of that crop. Estimates of the land required for feedstock production or the crops that are displaced are highly sensitive to estimates of future crop yields on both existing and converted land.

I. INTRODUCTION

What Is the Issue?

In recent years, concerns have been raised about the magnitude of land-use change that could be generated by scaling up biofuel feedstock production. The reliance on use of the U.S. agricultural land base to provide fuel as well as food, feed, and fiber will lead to increased competition for productive agricultural land, shifts in land use among different crops, and, in some cases, conversion of land from other uses into agricultural production.

Land-use change in this case refers to the conversion of land from some other use for the production of biofuel feedstock or for some portion of the production displaced by expanding biofuel feedstock production. The concern first arose during life-cycle analyses (LCA) of the greenhouse gas (GHG) emissions due to increased production of biofuels. Early LCA research for GHG accounting emphasized the importance of considering the underlying land-use changes that might be associated with increased feedstock production. Furthermore, because the climate change impact of GHG emissions depends on the aggregate of emissions worldwide, not the location of the emissions, land-related emissions around the world must be accounted for if they are induced by biofuels production. Initial estimates of potential

Measuring the Indirect Land-Use Change Associated ... 55

carbon release from global land-use changes suggested that so much carbon dioxide would be released under some land conversion scenarios that biofuel feedstocks would have to be produced on that land for hundreds of years as a way to compensate (Searchinger et al., 2008a; Fargione et al., 2008). These findings drew attention to the methods used to generate such numbers, and propelled the land-use issue into the forefront of the biofuels debate.

Economists have a long history of policy analysis in agriculture, though most land-use analyses are limited in scope to the impact of various domestic or international trade policies or technologies on existing domestic cropland resources. In contrast, studies attempting to quantify the broader land-use implications of biofuels production and biofuels policy, including competition among land-use sectors and market-induced repercussions worldwide, have appeared only in the last 2-3 years. To date, the most advanced efforts to integrate consideration of land-use-related GHG emissions into policy have been associated with California's Low Carbon Fuel Standard (LCFS), along with various efforts by member nations of the European Union (EU) to establish sustainability standards for biofuels, including the U.K.'s Renewable Transport Fuel Obligation (RTFO), and analyses associated with the U.S. Energy Independence and Security Act (EISA) of 2007.[1]

The regulatory requirements for GHG effects stipulated by the LCFS and EISA have greatly accelerated the development of tools necessary to perform life-cycle analyses of the GHG effects of biofuel production and to address questions related to land-use change induced by biofuels production. Such analyses require making projections about future values of domestic and international crop supply and demand, however, as well as broader assumptions regarding projected economic and population growth and development behavior. Analyses involving expectations about the future are inherently uncertain. Analytical methods exist for dealing with uncertainty in decisionmaking, but it is critical to understand the source of uncertainties in quantitative analysis to best determine how to manage them.

This Report

During the fiscal year 2010 appropriations process, the House Appropriations Committee directed "the Secretary of Agriculture through the Department of Agriculture's Economic Research Service, in conjunction with the Office of the Chief Economist, to do an independent study of significant indirect land-use changes for renewable fuels and the feedstocks used to

56 United States Department of Agriculture

produce them."[2] This study was conducted in response to that request. It summarizes the current state of knowledge of the drivers of land-use change and the role of biofuel production in affecting land-use change. It also details the analytical frameworks used to address land-use impacts of increased biofuels production, and the modeling procedures. The objective was to survey the literature in a neutral, objective way. There was no intention to suggest that USDA does or does not agree with any assumptions or model results.

Modeling the future land-use implications of major changes in global market conditions is fraught with many uncertainties, but the assumptions that are used to address these uncertainties significantly affect the results with respect to estimating the environmental impacts associated with changes. In accounting for GHG emissions, for instance, the results differ greatly depending on whether virgin forest or pasture is converted to increase the availability of cropland. Associated impacts on biodiversity and environmental quality from changes in nutrient, pesticide, and tillage use also will depend on the type of land that is converted.

Scientific methods for estimating the impacts of land conversion are improving to the point that estimates can be made if the type and location of the land-use change is known. Because the underlying stressor–the land-use change itself–is common across environmental dimensions; however, this study focuses on efforts to estimate the extent and location of land-use changes, not on efforts to estimate the resulting GHG or other environmental impacts of those changes.

Many studies of land-use change have been published in the past few years. The analytic approaches used in these studies have been expanded and refined over time, and the underlying assumptions used are now more clearly understood than in earlier studies. Some of the uncertainties associated with predicting indirect land-use change have been narrowed, while others are being examined to identify critical data gaps that are impeding development of more precise estimates.

Although most of the studies that are cited in the following review address the indirect land-use effects of corn-based ethanol production, the principles of land-use change remain the same for advanced biofuels production from other feedstocks. Corn ethanol results are used to illustrate the dynamics of land-use change because corn is currently the major biofuel feedstock in the U.S., the parameters of corn production are known, and models of global trade in corn are well established. Soybean use for biodiesel has also been studied, but the amount of soybean crop diverted to fuel production has been relatively small compared with that of corn. Recent breakthroughs in cellulosic conversion technologies promise a wider range of feedstocks for ethanol production and the potential for production of "drop in" fuels that would substitute directly for gasoline. In addition, bioenergy feedstocks may become important inputs in the generation of heat and electricity with minimal need for conversion to liquid fuel. [3] While the relative importance of particular drivers may change depending on feedstock, the underlying drivers of land-use change that are presented in this report will hold for any bioenergy feedstock (or any activity) that competes for agricultural land.

II. BIOFUELS, GHG EMISSIONS, AND LAND-USE CHANGE

What Is the Current State and Future Growth of Biofuels Production?

In 1925, Henry Ford told a reporter "The fuel of the future is going to come from fruit like that sumac out by the road, or from apples, weeds, sawdust—almost anything. There is fuel in every bit of vegetable matter that can be fermented. There's enough alcohol in one year's yield of an acre of potatoes to drive the machinery necessary to cultivate the fields for a hundred years." Ford initially designed his Model T to run on a mixture of alcohol and gasoline, but the advent of inexpensive petroleum products, together with Prohibition's strict limitations on ethanol production, redirected development of the automobile industry towards hydrocarbon, rather than carbohydrate, combustion. Despite a transitory resurgence during WWII, when oil was scarce, the use of ethanol as a motor fuel remained in the shadows until the 1970s.

In the last 30 years, however, several factors have reinvigorated interest in ethanol:

- The Arab oil embargo, which highlighted the Nation's economic vulnerability to oil imports from unstable regimes and the need to diversify U.S. energy sources.
- Increased demand for gasoline oxygenates, which were mandated by the Clean Air Act to decrease smog-related vehicular emissions, and the discovery of adverse health and environmental impacts associated

with other performance-related fuel additives such as lead, benzene, and MTBE.[4]

- The ability of ethanol production to stimulate agricultural markets, offering a potential market-based alternative to farm-support programs.
- The concern that GHG emissions from fossil fuel combustion are a key causal factor for climate change.
- The improved cost-effectiveness of ethanol production.

Moreover, the evolution of technology to produce biofuel from cellulose, which is in the demonstration stage and appears to be on the cusp of commercialization, has the potential to expand the spectrum of biomass raw materials that can be converted to biofuels.

To catalyze expansion of renewable fuels markets in the United States, Congress included in the Energy Policy Act of 2005 (EPAct) a Federal Renewable Fuel Standard (RFS) that mandated increased blending of renewable fuels into the U.S. fuel supply. In response to both increased interest in biofuels and increased concern about the potential GHG emissions associated with the production and use of conventional biofuels, the RFS was expanded and modified in 2007 to include a breakdown of fuel types, together with threshold levels of GHG emission reductions that the fuel types must meet to be considered qualifying renewable fuels under RFS II (Figure 1).

The sequential blending mandates, in conjunction with a tax credit that has been available to blenders who blend renewable fuels into gasoline or diesel, has been successful at accelerating development of the ethanol industry within the United States. Figure 2 shows the growth in U.S. corn and other starch-based ethanol production since 1992 as well as the forecast for growth of total ethanol production, from both starch and cellulose, to 2035 based on the latest long-term forecast from the Energy Information Administration (Annual Energy Outlook (AEO) 2010). The U.S. Department of Energy (DOE)-Energy Information Administration (EIA) projection of a brief plateau at roughly 15 billion gallons over the next decade reflects the limits placed on the volume of non-advanced ethanol that may qualify for credits under the RFS in the EISA; production levels eventually rise again due to mandated minimum levels of cellulosic biofuel under the RFS and projected improvements in the profitability of cellulosic ethanol.[5]

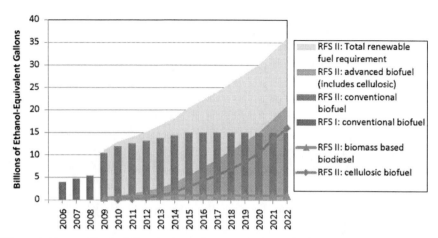

Figure 1. Renewable Fuel Standard requirements in the United States (billions of ethanol-equivalent gallons). Source: Renewable Fuels Association and Energy Information Administration.

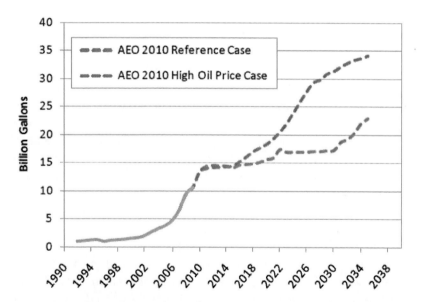

Figure 2. Historical and projected ethanol production in the U.S. (billion gallons). This graph illustrates only domestic ethanol production from starch and cellulose; it does not include projections of other biomass-based non-ethanol fuels, such as biodiesel. Source: Renewable Fuels Association and Energy Information Administration.

To meet demand for ethanol use, corn production in the United States is expected to increase from 12.9 billion bushels in 2009/10 to 14.0 billion

60 United States Department of Agriculture

bushels in 2015/16, with acreage planted for corn increasing from 86.4 million acres to 89.5 million acres. Over that period, yields for corn are anticipated to increase by about 5 percent, moving from 162.9 to 170.4 bushels per acre (USDA, Office of Chief Economist and World Agricultural Outlook Board, 2010).

Cellulosic Biofuels

The majority of ethanol investment continues to be targeted at proven technologies—generating ethanol from corn by converting starch, a simple sugar, to alcohol. However, the congressionally mandated RFS II objectives call for 16 billion gallons of cellulosic biofuels to be produced annually by 2022. Cellulosic biofuels are based on the conversion of cellulose, a complex sugar, to alcohol. Cellulosic technologies allow the conversion of biomass feedstocks such as stalks, leaves, grasses, and even trees into ethanol. AEO 2010 projects that the cellulosic biofuels requirements under RFS II will be met by cellulosic ethanol as well as by a portfolio of new biomass-to-liquids fuels such as Fischer-Tropsch liquids, renewable (or "green") diesels, and pyrolysis oils (AEO 2010).

Cellulosic conversion technologies for the production of ethanol offer significant benefits over grain-based production of ethanol, including a higher ethanol yield per acre from a diverse array of feedstocks, the use of perennial feedstocks that require less intensive management than annual grains, and a significant reduction in the demand for fossil fuels during processing.[6] Due to the variety of potential feedstocks for cellulosic processes, it could also be possible to locate cellulosic ethanol processing plants closer to high-demand areas, lessening the need to transport ethanol from the Midwest corn-growing regions to the more populated coasts. The potential benefits of cellulosic processing are increases in land-use efficiency, reduced carbon intensity per unit of energy content, gains in cost-effectiveness, and greater flexibility in finding feedstock production pathways with a smaller environmental footprint.

Obstacles to commercial deployment of cellulosic ethanol remain, such as high production cost and the need for separate pipelines and distribution systems, though technological breakthroughs have been reported. The EPA's final rule for the RFS II adjusted the 2010 cellulosic biofuel standard down from 100 million ethanol-equivalent gallons to 6.5 million ethanol-equivalent gallons.

In the face of ongoing obstacles, uncertainties remain regarding the speed with which cellulosic ethanol production will be able to scale up and the types of feedstocks that will be available for use. The potential environmental and GHG emissions impact of cellulosic biofuels will vary depending on the choice of feedstock. While the option of using "non-food" feedstocks has been touted as one of the advantages of cellulosic ethanol production, in fact only the use of waste streams as a biomass source truly eliminates competition with food production. Corn ethanol competes with food and feed supply chains for agricultural output, and cellulosic ethanol from many biomass sources will continue to compete with food and feed supply chains for agricultural inputs, most notably land. Therefore, while greater potential per-acre ethanol yields from biomass could keep land demand down relative to a similar scale of corn ethanol production, the dynamics of the land-use impacts addressed in this report will continue to hold at some scale should large tracts of land be converted to dedicated cellulosic energy crops, such as perennial grasses or short rotation woody crops.

Greenhouse Gas Emissions

Demand for biofuels stems from a desire to decrease use of fossil fuels, which would lower GHG emissions as well as increase energy security for the United States. Biofuel use recycles CO_2 (a major GHG) unlike the combustion of fossil fuels, which adds CO_2 to the atmosphere. However, the GHG footprints of different biofuel feedstocks are complex.[7] The GHG requirement associated with the RFS II set in motion an extensive effort to establish a regulatory LCA methodology that would determine which fuels and production processes satisfy the GHG threshold requirements.

Text Box 1: Life-Cycle Assessments

A life-cycle assessment (LCA) is a quantitative method of calculating the GHG balance of a biofuel (or other product) pathway, measured from production of raw materials through processing and transportation to end use. The recent expansion in the use of LCA for policy-based analysis, versus its conventional use as a product-based analysis, has prompted researchers to differentiate between "attributional" LCA analysis and "consequential" LCA analysis.

Attributional LCA analysis: Attributional LCA (aLCA) has historically

---

evaluated the environmental impact of a product through the quantification of input and output flows of the production system based on average data and fixing the system boundaries to those flows connected to the product under study. The conventional use of LCA—providing information on the impacts of the "average" unit of product over its production pathway—has been used to compare products to one another and to identify opportunities within the production system to reduce the impacts of producing that product (Brander et al., 2008). The research questions addressed by attributional LCA analyses generally do not require that indirect production impacts be included within the scope of analysis.

Consequential LCA analysis: Consequential LCA (cLCA) evaluates the aggregate environmental impact of a change in the level of output of a product. A cLCA approach is appropriate in the case of policy analysis such as that performed for the RFS II, where the question of interest is the GHG impact of scaling up biofuels production: "cLCA models the causal relationships originating from the decision to change the output of the product, and therefore seeks to inform policy makers on the broader impacts of policies which are intended to change levels of production" (Brander et al., 2008).

The relevant scope of a cLCA analysis is therefore broader than that found in aLCA; whereas aLCA performs its analysis taking the scale of the existing production system as fixed, cLCA is specifically interested in looking at the impacts of changing the aggregate scale of production. Such impacts include effects both inside and outside the life cycle of the product itself. Production processes are linked through competition for inputs and infrastructure, for instance, so changes in the scale of production of one good will induce changes in the production of other goods that compete for common inputs. Indirect impacts therefore become a key factor in consequential LCA analysis.

While the aLCA approach is generally static and based on fixed relationships between inputs and outputs at a given point of time, a cLCA approach looks at the impacts of changing relationships, allowing for ripple effects across sectors. Expanding the scope of life-cycle analysis to include market effects requires the integration of complicated economic models to represent relevant relationships between demand for inputs, prices, elasticities, and supply chains for products and co-products.

For that reason, some researchers caution that the results of consequential life-cycle analysis may be less precise than those of attributional life-cycle analysis (Brander et al, 2008), while others argue that, despite the additional uncertainty, the results are more comprehensive and complete (Schmidt, 2008).

Defining the Types of Land-Use Change Arising from Biofuels Production

Increased demand for land to produce crops for biofuel feedstocks involves two interrelated types of land-use change—"direct" and "indirect." **Direct land-use change** refers to the conversion of land from some other use directly into biofuel feedstock production. Estimating the magnitude of direct land-use change in the country or region producing biofuel is a matter of projecting where feedstocks of different types are likely to be grown and what pre-existing land use or uses are likely to be replaced (e.g., pasture land, existing crop land, forest land or idle or degraded cropland).

Land uses displaced by feedstock production may move outside the country or region where they originated. However, **Indirect land-use change** refers to the conversion of land to produce some portion of the production displaced by expanding biofuel feedstock production. Because the aggregate demand for goods that require land for production increases as a result of increased demand for biofuel feedstocks, that initial, direct change in land use for feedstock production could trigger a cascade of off-site, induced land-use conversions elsewhere as land uses are redistributed to satisfy demand. Prices are the mediating factor between the action of increased feedstock production for biofuels and the effect of distant land-use change.

Indirect land-use changes can occur domestically or globally. For instance, increased corn production in the Midwest to supply ethanol production could induce the withdrawal of land from the Conservation Reserve Program (CRP) to grow wheat that has been displaced by the expansion of corn production for feedstocks. The conversion of CRP land to wheat is considered a domestic indirect land-use change associated with the initial increase in demand for corn for ethanol production. Sometimes, the trail from cause to effect is quite long, with many steps along the way. In the ethanol example, increased corn demand reduced soybean planting in the U.S., which reduced soy exports, which increased world soybean prices, which increased soybean planting in Brazil on land previously used to pasture beef, which increased beef prices, which increased incentives to clear Brazilian forests to increase pasture (Zilberman et al., 2010). The LCA accounting for GHG emissions is affected by the selected scope of the analysis. Changes in output in markets other than biofuel feedstock and ethanol production will occur as a result of price changes induced by increased biofuel production, and those changes in output may also have emissions implications.

64　　United States Department of Agriculture

Although researchers agree upon the basic concepts of direct and indirect land-use change, common usage of the terms has added confusion to the debate. "Indirect" and "international" are often used interchangeably. As noted earlier, the global land-use changes that are induced by U.S. biofuel feedstock production may be significant, but there are many drivers of international land-use change that will affect conditions abroad regardless of domestic biofuels policies.

Drivers of Land-Use Change and Implications for the Agricultural Sector

Increased biofuel production has had and will have effects on land use in the U.S. and the rest of the world as will any change that increases competition for agricultural land. Estimating the magnitude and location of that impact, however, is extremely challenging. In particular, estimating the indirect land-use changes that are attributable to biofuels production or policy is subject to a great deal of uncertainty. The extent of indirect land-use change is a market-mediated response that plays out internationally through global markets via changes in relative land values and commodity prices. The extent to which displaced production is replaced will depend on elasticities of supply and demand,[8] and where that production will migrate is subject to constraints imposed by national and international policies related to trade, energy, land use, forest management, etc. (Marshall, 2009). Furthermore, the difficulty of attributing off-site land-use changes to changes in feedstock production is exacerbated by the effects of simultaneous drivers of land-use change such as population change, income change and the associated changes in demand for goods such as meat and housing space, and diverse policies related to land-use management, development, trade, and other countries' biofuels policies – each of which affects the competitive advantage of different lands in different uses.

Landowners will allocate land among competing uses based on the expected net benefits of those uses, and those benefits will vary for each use depending on land quality and location. A landowner seeking to maximize profits will allocate a land parcel to the use that yields the highest expected economic return after the costs of conversion, which can include changes in machinery investments and management practices. Relative expected returns change with market conditions (commodity prices, production costs, population growth, consumer tastes, international trade, and other factors affecting the demand for land in different uses), technological advancements,

Measuring the Indirect Land-Use Change Associated ... 65

and weather. The level of uncertainty surrounding future conditions will affect a landowner's assessment of expected benefits and costs.

Drivers of land-use change in the United States can be roughly categorized into those that may encourage either expansion or contraction of cropland acreage depending on market conditions, those that primarily encourage expansion, and those that primarily induce contraction of cropland acreage.

Examples of land-use change drivers that can induce expansion or contraction of cropland acreage:

- **Crop prices** can influence the amount of land planted to various crops because they affect the relative profitability of crops and farm income.
- **Changes in input costs** can also affect profitability. For example, use of land for cropping often requires application of fertilizers. Fluctuations in fertilizer costs can change the returns to cropping relative to other land uses. They can also affect decisions to grow particular crops because some crops require greater application of agricultural chemicals relative to other crops.
- **Technological change**, such as the introduction of yield-increasing crop varieties can increase the net benefits on a given piece of land and may reduce the demand for expansion of acreage. Some innovations, such as drought resistant seeds or pressurized irrigation systems, can induce expansion onto lands with lower quality soils.

Examples of land-use change drivers that primarily encourage expansion include:

- **Agricultural policies that increase expected returns and reduce the inherent risk associated with agricultural production** can increase the relative benefits of that land use. Such policies include commodity support programs and the Federal crop insurance program.
- **Energy policies, such as the Renewable Fuel Standard, that stimulate demand for existing commodities or create a market for new agricultural products,** such as perennial energy crops, can change the relative benefits of land use.

Examples of land-use change drivers primarily inducing contraction in cropland acreage:

- Urbanization and pressure for commercial, residential, and industrial development can increase the demand for land and reduce the relative benefits of keeping land in agricultural uses.
- Conservation policies that mitigate or reverse the environmental impacts of conversion can increase the benefits of retiring land from agricultural production. Since 1985, the CRP has been the largest driver of cropland changes (Lubowski et al., 2006).
- Conservation policies that protect vulnerable natural resources from conversion to cropland or higher intensity uses can offer benefits to farmers. Such policies include the Grassland Reserve Program and the Wetland Reserve Program.

Because future patterns of land use will depend on complex interactions among all of these forces, projecting future landscapes, and the degree to which they will be impacted by changes in a single driver, is very difficult. Broadening and increasing the portfolio of food, feed, fiber, and energy products from domestic agriculture will increase the amount of land required to produce those products and change domestic patterns of production. The response in terms of domestic land use will be sensitive to the wide array of factors described above.

Global Perspective

U.S. agriculture is an integral part of global commodity markets, and changes in domestic production will impact how global demands for food, feed, and fiber are met. If competition for U.S. land reduces the availability of commodities for export, for instance, global markets will create the incentive, through higher prices, for those commodities to be produced elsewhere. Estimating the patterns of land-use change that will enable that increased production internationally is the crux of the indirect land-use change issue. Because agricultural expansion is one of the key drivers of land-use change and deforestation in many developing countries, there is little question that some international land-use change will result from domestic shifts in production. The debate intensifies, however, around the question of extent and location of those changes. Policies within a country that affect migration (population shifts) and associated land-use change can include road construction, colonization policies, agricultural subsidies, and tax incentives (Angelsen and Kaimowitz, 1999). While estimating future land-use change in a single, relatively data-rich country such as the United States is difficult, estimating waves of land-use change that propagate internationally through complex global commodity markets is even more challenging.

III. METHODOLOGIES FOR ESTIMATING THE FUTURE LAND-USE CHANGE ASSOCIATED WITH BIOFUELS PRODUCTION

As discussed previously, land-use change is influenced by a number of factors, and the interactions between these factors are complex. Because experiments cannot be carried out to determine the consequences of policy or market changes, alternative means must be used. Mathematical models that attempt to estimate and test the interrelationships between factors are often used to quantify these relationships.

A quantitative economic model is a mathematical representation of how agents in a system behave under a set of hypothesized relationships informed by both theory and empirical evidence. A model serves as a proxy for what one cannot actually observe. In agricultural sector models that attempt to model land-use change, the agents are often producers of agricultural commodities and livestock, biofuel producers, and Government policy makers. A model is used to indicate, numerically, how the aggregate behavior of the agents will change if some facet of the production environment changes, such as through an increase in biofuel production mandates.

Because one cannot have perfect foresight about variables related to future weather, policy, and demand conditions, or many other factors that govern land-use decisions, the output of a model should not be taken as an immutable prediction of how the future will unfold. To isolate the effects of the variable of interest—in this case biofuel production—from the many other potential sources of uncertainty when projecting future land-change, modeling efforts usually measure the land-use impacts attributed to biofuel feedstock production as the difference between two future modeled, or projected, scenarios. Between those scenarios—a baseline projection and a "scaled up biofuels" projection—the only variable that differs is the change in biofuel production volume.

68 United States Department of Agriculture

The resulting difference in land use is therefore attributed to the biofuel policy mandating increased production. For example, the USDA World Agricultural Outlook Board coordinates a multiagency process that projects agricultural supply and demand 10 years out based on explicit assumptions about world markets, yields, and agricultural trade and environmental policies. This baseline is used in several studies as a projection against which to assess policy scenarios.

Because the two future scenarios share a set of common assumptions about exogenous future land-use drivers such as GDP and weather pattern, the estimate of land-use change derived this way does not reflect additional uncertainty surrounding what weather patterns are likely to be. Changes to these common data or assumptions, however, will change the resulting estimates, and, depending on the sensitivity of the model to those elements, may have a significant influence on the output. An important component of such analyses are therefore "sensitivity analyses" to determine whether the land-use change estimates respond significantly as input parameters and data are changed. These analytic frameworks can be used to assess the sensitivity of results to assumptions such as yield growth and future prices. It is important for any modeling study to be transparent about the assumptions used and the degree of confidence in the accuracy of input data.

Text Box 2: Model Frameworks

The first step in assessing quantitative and qualitative land-use impacts associated with a policy that will affect agriculture and forestry production decisions is choosing an analytical framework. There are several types of quantitative models that are used, and none can claim to be the ideal tool for assessing land-use change; each type of model has advantages and limitations.

Partial Equilibrium Models

Partial equilibrium (PE) models feature a detailed representation of agricultural (and/or forestry) production for a country or region. They typically utilize observed data to determine the amount of inputs required to produce a unit of a given product.

The representation of production and land use can be highly detailed, allowing for variations in crop rotation, tillage, fertilizer application, and

Measuring the Indirect Land-Use Change Associated ...

other variations in farm-level production decisions. Such models are "partial" in the sense that they focus on a subset of economic sectors and do not link explicitly to other sectors of the economy. Depending on the ways in which crop, livestock and forestry sectors are modeled, substitution between inputs may or may not be allowed. Partial equilibrium models can track movement of and competition for inputs such as labor, water, energy and fertilizer within the modeled sector(s), but economy-wide competition is beyond the scope of most PE models. Export and import of agricultural products and inputs are usually modeled in a relatively simple manner.

Partial equilibrium models that have been used to assess land-use change in agriculture include the Regional Environmental and Agricultural Production model (REAP), maintained by the USDA's ERS, and the Forestry and Agricultural Sector Optimization Model (FASOM), at Texas A&M University. Both of these models focus on domestic crop production. The Food and Agricultural Policy Research Institute (FAPRI) maintains a family of econometric agricultural models that are coordinated to produce domestic and international agricultural production projections. The FAPRI models address only the agricultural sector and in aggregate can be described as a non-spatial, partial equilibrium agricultural sector model that covers both domestic and international production.

General Equilibrium Models

Computable general equilibrium (CGE) models are a class of economic models that use observed economic data to estimate how an economy might react to changes in policy, technology, or other external factors. CGE models attempt to portray an entire economic system (national or global) by accounting for interactions between all productive sectors, labor, flow of goods and capital between sectors (or countries), and government policies. The tradeoff for the expansion of modeling scope is often a loss in modeling detail for particular sectors such as agriculture. Furthermore, most existing global and regional CGE frameworks are not structured to model land-use alternatives and the associated emissions sources and mitigation opportunities. This work has been hindered by a lack of data – specifically, consistent global land resource and non-CO_2 GHG emissions databases linked to underlying economic activity and GHG emissions and sequestration drivers (Hertel et al. 2009).

Examples of CGE models used to evaluate land-use change impacts in agriculture are the Global Trade Analysis Project (GTAP), housed at

70 United States Department of Agriculture

Purdue University, and the Integrated Model to Assess the Global Environment (IMAGE) of the Netherlands Environmental Assessment Agency.

While CGE models feature a full accounting of the economic flows between production sectors, they have not been used widely to model land-use change at the regional level. The recent research emphasis on the importance of land and land constraints in GHG analysis of policy, products, and trade is changing that. Recent modifications to the GTAP model have included creation of a land-use module that allows GTAP-Bio to represent the global competition for land among land-use sectors (Golub et al., 2009).

Dynamic vs. Static Models

A static model addresses the impacts on production for a single year. While there are dynamic elements to agricultural production that are a function of the previous year's crop, such as commodity storage and crop yields, these elements can often be handled adequately in a static model through the use of averages or capitalized values. Dynamic optimization models explicitly model the evolution of the production environment over a range of time and can attempt to optimize over a multi-year pathway.[1] Dynamic models, for instance, are frequently used to model the forestry sector, with its long rotation times and fractional annual harvest.

Static models generally solve much faster than dynamic models and are useful in simulating a large number of policy shocks as uncertain inputs are varied. Dynamic models can be much more computation-intensive and are more sophisticated in analyzing transition effects and the effects on later periods of a decision made today under different states of the world.

A robust projection of the domestic and international land-use implications of biofuel feedstock production requires an integrated modeling system that is capable of providing answers to a lengthy list of complex and interrelated questions:

- How much feedstock will be required to meet projected biofuel demand?
- How much land will be required to produce that much feedstock?
- Where will land for feedstock production come from?
  - o What will be the methods and costs of feedstock production?

Measuring the Indirect Land-Use Change Associated ...

- o What are the available sources of land, beyond existing cropland, for feedstock production?
- o How competitive will feedstock production be with existing land uses such as other crops, pasture, and forestry?
- o What additional obstacles exist to farmer adoption of feedstock production?
- o What are the environmental implications of changing land uses?
- What will be the impacts of changing patterns of domestic production on international commodity prices?
- Which countries will respond to changing international prices with changes in agricultural production patterns?
- What lands within those countries will be affected by changing patterns of production? If output is increased, what types of land will be converted into production?

No single model currently exists that can answer such a broad list of questions for feedstock production in the United States. Analyses at this scale generally tie together models that can answer the domestic questions related to production and macroeconomic impacts with other models that evaluate how international markets respond to regional changes in production and macroeconomic impacts. While such economic models have been under development for decades, the introduction of an explicit land-use component into the economic framework is a relatively recent addition. The link to land use in these models has been complicated by a lack of sufficient, consistent, and comparable data regarding existing land uses and historical patterns of land-use change for the United States and worldwide.

The sophisticated integration of models required to estimate biofuels-related land-use change highlights the complexity of the forecasting task. Models capturing the land-use change dynamics described earlier must make projections about future values of parameters ranging from farm production practices for crops that are not yet grown commercially, to expected growth in existing crop productivity, to projected responsiveness of world markets to changes in crop prices under specific global economic growth scenarios, and to enforcement of land-use policies in other countries.

Future values for parameters such as these are not yet known, so judgments and assumptions must be made as to the likely values these uncertain data will take. Each assumption, whether made explicitly or implicitly in the structure and data of the model, will influence the outcome to some degree, though the extent to which they influence land use results varies, with some parameters generating significantly more variability across their plausible ranges than others.

There is a long history of research on analytical methods for exploring and illustrating uncertainty in analytical contexts such as this one; model-based analysis is particularly well-suited to indepth exploration of how outcomes vary as the range of possible input values and value combinations is explored. The results of such analyses are used to refine outcome estimates over time in multiple ways:

- Identifying the scope of analysis that encompasses and elaborates on the most significant variables in generating outcomes and outcome variability;
- Reducing uncertainty surrounding those variables where uncertainty is a function of missing or coarse information; and
- .Managing or bounding the uncertainty around those variables that are inherently uncertain because they are unobservable, such as future projections.

The evolution of the indirect land-use change modeling efforts has made progress along each of these pathways. Before describing the status and evolution of that analytical effort, the following section briefly reviews the major sources of uncertainty in this analytical context.

IV. SOURCES OF UNCERTAINTY IN MODEL METHODOLOGIES AND ASSUMPTIONS

The sources of uncertainty in modeling the indirect land-use change associated with biofuels production have been roughly grouped into five categories:

1) The uncertainty associated with the demand for land for feedstock production;
2) The uncertainty associated with the supply of land in the United States;
3) The uncertainty associated with responses of domestic and international markets to feedstock production;

Measuring the Indirect Land-Use Change Associated ... 73

4) The uncertainty associated with the magnitude of land-use changes in those countries responding to international price signals; and
5) Other uncertainties in modeling indirect land-use change.

Assumptions made about the uncertain future not only affect the model results directly but also determine the baseline from which change is measured. Although a few models measure change from current conditions (attributing the total change to biofuel feedstock production), most project the likely conditions at some future date with and without the activity or policy of interest. The difference between the baseline and scenario analysis is meant to capture the policy effect. For example, the USDA World Agricultural Outlook Board coordinates a multiagency process that projects agricultural supply and demand 10 years out based on explicit assumptions about world markets, yields, and agricultural trade and environmental policies. This baseline is used in several studies to assess policy scenarios and the sensitivities of the results to the underlying assumptions embedded in the model.

The sections that follow explore some of the assumptions that are significant in generating variability around estimates of land-use change associated with future biofuel feedstock production.

Uncertainty Associated with the Demand for Land for Feedstock Production

As demand for land for biofuel production rises, competition for that land will intensify and the effects of indirect land conversion will likely grow as well. Several factors influence direct feedstock demand for land, including feedstock yield assumptions and substitutability of biofuel coproducts for other products that require land as an input. Any bioenergy feedstock that has the potential to be grown on agricultural land can be considered a competitor for that land. The type of land on which it will be grown and the current use that it will displace will depend on the resource requirements of the feedstock and the relative profitability of that feedstock compared to the current use. One of the most important determinants of profitability is the yield that can be produced per acre.

Yields
A crop variety has a "yield potential" that represents the expected yield if all conditions are perfect. The yield potential can be changed through research

74 United States Department of Agriculture

(or mutation). Actual yields are an outcome of environmental conditions and the choices made by the producer about the use of agricultural practices and inputs such as tillage, irrigation, and fertilizer. If crop yield per acre increases through more intensive management or new crop varieties, then less land is needed to grow a particular amount of that crop. This is true for bioenergy feedstocks and any crops that they displace. Estimates of the land required for feedstock production are therefore highly sensitive to estimates of future crop yields on both existing and newly converted land.

Estimating future crop yields on newly converted lands is complicated by uncertainty about the productivity of land that has not yet been converted. Unfortunately, there is little empirical evidence to guide modelers in selecting the appropriate value for estimating the productivity of converted land. In most regions, existing crops are already on the most productive agriculture land, so yields on newly converted lands would likely be lower than on existing cropland.[9] New crop varieties or more intensive input use may mitigate such yield losses. The expected productivity of new land in agriculture is a major determinant of how much new land will be required to accommodate increased demand for biofuel feedstock or any use that competes for land.

The uncertainty about yield on land converted today to a new use is compounded when yields are projected over time. The yield potential for most crops has grown over time due to public and private investments in plant breeding, biotechnology, and other crop improvements (Figure 3). More intensive use of existing technologies and the adoption of new technologies also can enhance productivity. Yields, for instance, can be increased in response to higher prices by more intensive use of inputs such as fertilizer (Keeney and Hertel, 2009). Greater input use or more intensive management, however, may have environmental consequences. Increased nitrogen application may result in increased direct N2O emissions, and more intensive farming practices may result in increased erosion and decreased soil carbon sequestration. If one of the underlying motivations for concern about indirect land use is concern about GHG emissions, then such potential tradeoffs between use of land and use of other inputs to increase production must be acknowledged and incorporated into a comprehensive analysis. This concern extends to cellulosic feedstocks; although cellulosic feedstocks such as perennial grasses can be grown with fewer inputs than a crop such as corn, yields can be improved through the use of fertilizer, and added nutrients will be applied if increased revenues outweigh the costs.

Pest-resistant and herbicide-tolerant biotechnology-derived varieties of corn, cotton, and soybeans have been adopted extensively by farmers in the

United States, but acceptance has not been as widespread in other parts of the world. Current biotechnology research is focused on traits such as drought tolerance that would increase yield potential for many crops. Biotechnology research will be employed in the development of bioenergy feedstocks as well as traditional agricultural commodities. Whether the recent trend in yield growth will continue unchanged, increase, or decrease is a matter of agronomic limits and research investments (Heisey, 2009). Estimates of the rate of yield growth may also differ across crops and locations. Productivity performance across countries and regions has not been uniform (Fuglie, 2010).

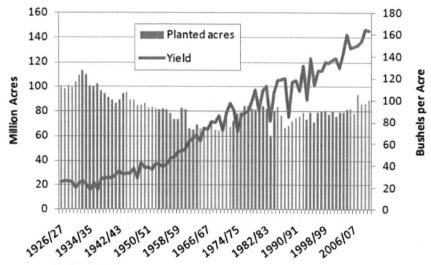

Figure 3. Yield increases for corn over time in the United States.

Source: USDA, Office of the Chief Economist and World Agricultural Outlook Board, 2010.

Text Box 3: Scientific Advances, Economic Factors, and Biological Limits Determine Crop Yields Over Very Long Periods of Time

The very long-run trajectory of yields for a major field crop like corn usually begins with a phase in which yields are flat. Yields then increase over a long time with the application of science to agriculture.

Finally, yields might reach some theoretical maximum based on the plant's capacity to capture available sunlight, efficiently convert it to biomass, and partition that biomass into grain. The resulting simple S-shaped graph may conceal a number of important factors. First, flat yields may not always signal periods of technological stagnation. Mechanical technology may have lowered the labor inputs into land preparation, cultivation, or harvesting. Different crop varieties may have maintained yields that would have otherwise deteriorated because of pests and diseases, or allowed expansion of commercial crop production into different growing zones. Second, the yield trajectory over time and maximum attainable yield nationwide will differ from results on small-scale experimental plots. It takes time for new corn technology to move from scientific plots to commercial application by farmers, and additional time may be required to adapt corn technology to less favorable growing regions. Furthermore, maximum yields on experimental plots will differ from yields in farmers' fields because a greater number of detrimental factors can be controlled on small experimental plots, and farmers must consider economic costs and benefits, which are less relevant in trials aimed at maximizing yields.

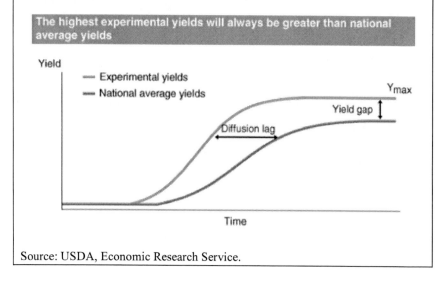

Source: USDA, Economic Research Service.

Coproducts from Biofuels Production May Mitigate Land-Use Pressures:

Because ethanol diverts corn from the livestock sector, it is often perceived to be a competitor, rather than a supplier, in the livestock feed market. Ethanol dry mill production produces a coproduct called distillers' dried grains (DDGs), which can substitute for corn as feed, thereby reducing the amount of corn required by the livestock sector. The greater the substitutability between DDGs and corn, the fewer total cropland acres will be needed to supply both the ethanol and livestock sectors.

Dietary considerations place constraints on the amount of DDGs that can be used in livestock diets. Furthermore, marketing challenges remain due to variability in nutrient composition, issues related to product storage and transport, and food safety concerns (Malcolm et al., 2009). Nevertheless, research currently being conducted on how animal feed rations can be modified to make use of DDGs will lead to an improved understanding of how and to what extent DDGs can substitute for feed products, thereby offsetting some of the increased land demand necessary for corn production.

Uncertainty Associated with the Supply of Land in the United States

The amount of pressure placed on land internationally will depend in part on how much of the land needed for biofuel production is met through an expansion of agricultural land in the United States. Agricultural production models for the United States have a long history of estimating the movement of land allocation among different crops on existing cropland based on changes in expected return. Land conversion between broad land-use categories, however, such as from forest to pastureland or cropland, can be both costly and irreversible and is therefore driven by longer-term economic factors. For example, Midwest farmers can readily move cropland between corn and soybeans when the relative profitability of those crops change. In contrast, the conversion of forest acreage to cropland represents a long-term change with high transition costs: such a decision must consider the relative profitability between agricultural and forestry commodities for many years into the future.

The uncertainty of the future values of prices and costs will affect an owner's expectation of long-term profits from any alternative. In addition, the partial irreversibility of some land-use changes gives rise to option values that increase the incentive to keep land in its current use (Stavins, 1999).[10] Using an option model, Song, Zhao, and Swinton (2009) found that landowners would be more reluctant to convert land from annual to perennial crops. The irreversibility of some land-use changes coupled with uncertainty about future economic returns can act as a brake on land conversion. Farmers hesitate to

78　　　　　　United States Department of Agriculture

convert land because they value the "option" or "flexibility" for future land-use decisions that is preserved when land is maintained in its current use. Isik and Yang (2004) also found that such option values play a significant role in farmer decisions to retire land and that they reduce the probability of farmer participation in land retirement programs.

Other factors may significantly affect land conversion decisions in a particular area or country, such as national or regional conservation and preservation policies and programs.

Uncertainty Associated with the Responses of Domestic and International Markets to Feedstock Production

If biofuel feedstock production disrupts domestic production of other crops through competition for land or other inputs, market prices of existing crops will be affected, which will send signals internationally that may result in changes to production patterns abroad. Increased commodity prices in 2007 sparked a debate about the effects of corn-based ethanol production on consumer food prices. The "food versus fuel" discussion demonstrated the lack of consensus on the extent to which use of a portion of the Nation's corn for ethanol production impacts the prices of finished food products, as well as the prices and export availability of the ethanol feedstock and other commodities (Trostle, 2008; Leibtag, 2008). This price signal, however, is the critical link between the behavior of the domestic agricultural sector and the induced land-use response projected for other agricultural exporters.

Given the size of U.S. agriculture's influence on international markets, producers and consumers overseas regularly react to market or policy-related events in the United States. Domestic adjustments in production and crop mix may change incentives for foreign producers to supply their own markets or the world market, by bringing new land into agricultural production. The larger the impact of domestic biofuels production on price and availability of exports (of both the feedstock and, through land competition, all other commodities), the larger the international land-use repercussions of that domestic biofuel production are likely to be. The strength of the signal sent to markets around the world is therefore highly dependent on the domestic land supply issue already described.

The nature and magnitude of the international market response to increased prices hinges on both the production and consumption response. The consumption response is reflected in estimates of the price elasticity of demand for agricultural commodities. This variable captures what happens to world demand for food and feed crops, including crops fed to animals for meat/dairy production, when their prices increase on the world market. Assumptions of inelastic demand, where food and feed consumptions are relatively unresponsive to price, lead to larger estimates of land use impact than when it is assumed that food consumption will decline sharply as food and feed prices rise; a relatively more elastic demand for agricultural commodity good production means that in aggregate, fuel production on agricultural land "crowds out" food production to a certain extent. There are consumer welfare implications associated with the increased prices and reduced consumption; Roberts and Schelenker (2010) econometrically estimate commodity demand elasticity and the consumer welfare losses associated with reduced consumption due to biofuels-driven price increases.

Predictions of international production responses to domestic market changes often rely on one of two assumptions concerning the economic mechanism by which goods are traded between countries. The first assumption asserts that trade generally occurs in response to a single world price that results from the production and consumption decisions of individuals around the world. Thus, after controlling for transportation costs and border-related barriers, only those producers that can profitably sell at the world price will participate in the world market, irrespective of their location or their past relationships with buyers.

The second assumption, first proposed by Armington (1969), contends that, in addition to prices, bilateral relationships between traders do matter and that consumers distinguish from which country their goods originate. Under the Armington assumption, an analyst must rely on a quantitative estimate of the strength of the bilateral trade relationship. Though models may be limited to a single assumption about trade structure, the two frameworks are not mutually exclusive; a realistic trade response may be best represented using an Armington assumption in the short term, but allowing for evolution toward a "single world price" structure in the long term.

Uncertainty Associated with the Magnitude of Land-Use Change in Those Countries Responding to International Price Signals

Each country participating in the global market has its own characteristics with respect to current land use, land quality, resource availability, legal framework, demographics, infrastructure, international relations and economic conditions. All of these factors must be considered when analyzing the effects of changes in agricultural prices on the allocation of land in a particular country.

Using trade analysis to estimate the responses of international markets in such cases generally involves predicting changes in exports and imports of specific countries or regions under the new production scenario. The next step in tracking indirect land-use change is to associate those changes in production with the underlying land-use changes in each country that would be required to meet new export and import patterns. This requires a calculation of how much land would be required to meet the changing export/import patterns and a projection of where that land would come from. The amount of land converted into agriculture, and its original coverage (e.g., forests, pasture, or idle land), can critically determine the one-time release of carbon emissions attributable to land-conversion impacts. The challenge, therefore, is identifying where land conversion will occur, and which type of land will get converted.

All of the land-use drivers described here are relevant in the international context, but in varying degrees. As in the U.S., land prices, development pressure, access to markets and transportation costs, and opportunity costs associated with alternative uses of land need to be included in the model, along with information on the effectiveness of national, regional and local land-use policies such as protected areas and conversion set asides. Lack of sufficient data makes it especially difficult to model or project future land-use change in foreign countries. Researchers must therefore make a number of simplifying assumptions about how land-use change within the country is likely to occur in the future under business as usual (the baseline) as well as how that baseline scenario will change under the biofuels feedstock production scenario. Cross-country assumptions that U.S. biofuel policies will have the same type of impact on land use within each country may result in misleading conclusions because each country has a unique set of resource endowments, institutions, trade relationships, and economic drivers (Angelsen and Kaimowitz, 1999).

Other Uncertainties in Modeling Indirect Land-Use Change

Many forces of varying magnitude will affect future conditions that, in turn, will affect the extent of indirect land-use change. For example, extreme weather events, natural disasters, political instability or conflict, and technological breakthroughs all have the potential to alter global supply and demand for agricultural products. One of the biggest determinants for biofuel feedstock production is the energy market, which includes the supply and demand for fossil fuel as well as all renewable energy sources.

Some uncertainties in modeling indirect land-use changes that will not be explicitly addressed in this report are associated with global energy markets and the interactions between energy demand and population growth, economic growth, and income generation more broadly. Biofuel feedstock production depends on the demand for renewable energy. Higher gasoline prices in recent years decreased the U.S. public's demand for energy in general and increased its demand for alternative energy sources. As was observed after the "oil crisis" of 1973, however, the public interest in energy efficiency and alternative energy can fade quickly when fossil fuel prices fall.

The long-term viability of renewable energy markets will depend on renewable energy being cost competitive with fossil fuels. Large investments are being made to lower biofuel production costs. However, it has been noted that relatively low energy prices can have the unintended effect of increasing energy use (the "rebound effect") and potentially increasing GHG emissions (Alfredsson, 2004; Beckman et al., forthcoming 2011).

Assumptions about future energy prices, including the impacts of such prices on economic growth and income generation, are therefore critical variables within the models that are used to estimate indirect land-use change. A related issue will be the direct competition in the provision of biofuels if domestic demand sufficiently increases. It is predicted that countries such as Brazil and Indonesia are likely to convert more land to biofuel production to supply U.S. renewable fuel needs (Fargione et al., 2008).

Another uncertainty associated with modeling energy markets that include both fossil and renewable fuels is predicting technological breakthroughs that can affect either supply or demand for one or more energy sources. Large investments are currently being made to improve cellulosic conversion so as to widen the range of bioenergy feedstocks that can profitably be used as fuel. However, research is also underway to develop technologies to process coal, tar sands, and oil shale with reduced emissions. Hydrogen power may be on the horizon, which, depending on the price, would affect demand for both

82 United States Department of Agriculture

bioenergy crops and fossil fuels. On the other hand, improvements in feedstock to fuel conversion efficiencies would increase the relative advantages of biomass-based fuels. Since such technological breakthroughs cannot be predicted with certainty, many models include assumptions about innovation rates that lower production costs over time. As with the crop yield increases described earlier, innovation depends on research investments that may or may not pay off. No assumption about a steady rate of technological improvement can capture the impacts of a "game-changing" breakthrough.

V. MODELING EFFORTS AND RESULTS TO DATE

Interest in measuring the scale of indirect land-use change attributable to biofuel production largely arose from concern about the potential GHG implications of such conversions. In January 2008, Dr. Alex Farrell, then director of the Transportation Sustainability Research Center at UC Berkeley, presented to the California Air Resources Board (CARB) Low Carbon Fuel Standard (LCFS) Program an illustration of potential indirect GHG impacts from biofuels production in the U.S. that would arise from indirect land-use change in other countries. That analysis, which was the first to attract major interest in the subject, presented what Farrell described as "crude upper limit estimates" on the GHG emissions associated with land-use change for biofuels production (Farrell and O'Hare, 2008). To get his extreme "worst case" scenarios, he made the assumption that an acre of bioenergy feedstock production will result in an acre of land conversion and coupled it with the assumption that the land lost will come from the highest carbon land use available for conversion. This resulted in estimates that an acre of corn production for ethanol in the United States would ultimately lead to an acre of tropical rainforest conversion, at conversion costs that go as high as 826 gCO_2e/MJ of energy from corn ethanol.[11] Less carbon-intensive scenarios where the analysis assumed that lower-carbon temperate grasslands, rather than tropical forests, are converted for corn production (while maintaining the 1:1 acreage conversion assumption) result in significantly lower estimates of 140 gCO_2e/MJ.

The evolution of estimation procedures since then has largely focused on identifying and improving the precision with which key variables—such as anticipated future crop yields and source, carbon-intensity, and productivity of new land brought into production internationally—can be represented in modeling efforts. In particular, the integration of land use into traditional economic models of production and trade, and the accompanying capacity to estimate the indirect land-use impacts of domestic production, has rapidly increased in sophistication in response to CARB and EPA's regulatory requirement to measure the GHG impact of biofuel production. Table 1 presents the results from several recent modeling efforts that estimate the effects of ethanol production on global land use. These studies attempt to quantify the market response in the United States and in other countries to increases in commodity prices due to increases in biofuel production. These studies also quantify the GHG emissions from these market responses and attribute these emissions to biofuel production. The table is not meant to be comprehensive but shows a selected range of estimates. Other models, such as MIT's Emissions Prediction and Policy Analysis model, have also been used to examine indirect land-use change impacts (Gurgel et al., 2007; Melillo et al., 2009).

While it is conceptually useful to distinguish between indirect and direct land use impacts in discussing the impacts of biofuel policy, modeling efforts often present an aggregate land-use impact estimate and do not attempt to distinguish between the two. Modeling frameworks may not be spatially explicit enough to directly associate increased commodity production with specific parcels of land, so they are unable to specify whether converted land goes to feedstock production (direct impact) or to other crops displaced by feedstocks (indirect impact). As specified under EISA, both types of impacts are critical components in determining the full land-use impact of a biofuel policy, so the aggregate estimate is an appropriate indicator for a comprehensive analysis of the land-use impacts.

The aggregate land-use impact estimates shown in Table 1 vary for a number of reasons. Some of the estimates are derived using modeling structures with different ways of representing relationships and different boundaries of analysis. Furthermore, these models are often based on, or incorporate, differences in assumptions about the many variables described earlier that can affect land-use estimates. Even where a path of values over time for a particular variable has a relatively narrow band of uncertainty, researchers may make different assumptions about the year in which the estimates are derived. The year of comparison is important in determining what technologies are assumed to be in place, such as those governing crop yields and ethanol yields per unit of feedstock. The section that follows describes in more detail the specific difference among the research efforts illustrated and how they lead to varying estimates of land-use impact.

Table 1. Estimates of land-use change and emissions related to land-use change associated with biofuel production

Study	Modeling framework	Increase in ethanol production	Change in Global Land Use	Change in Global Land Use	CO2 equivalent emissions
		Billion gallons per year	Million acres	Million acres per bil. Gal per year	Grams CO2e per MJ of ethanol per year 1/
Searchinger et al., 2008a 2/	FAPRI/ CARD	14.8	26.73	1.81	104
Fabiosa et al., 2009 3/	FAPRI/ CARD	1.174	1.923	1.638	na
California (CARB) 2009	GTAP	13.25	9.62	0.726	30
EPA proposed rule for RFS II (corn ethanol 4/)	FASOM/ FAPRI/ GREET	2.6 (12.4 to 15.0) (p. 422, RIA)	4.4 (table 2.9-3, RIA)	1.692	60.4
EPA Final Rule for RFS II (Corn Ethanol 4/)	FASOM/ FAPRI/ GREET	2.7 (12.3 to 15.0) (p. 311, RIA)	1.95 (table 2.4-29, RIA)	.722	30.3
Hertel et al., 2010	GTAP-BIO	15	9.4	.628	27
Tyner et al. 2010 5/	GTAP (scenario 3)				
2001 to 2006		3.085	1.155	.374	17.3
2006 to 7 BG		2.145	.577	.269	12.9
7 to 9 BG		2	.581	.291	13.4

Study	Modeling framework	Increase in ethanol production	Change in Global Land Use	Change in Global Land Use	CO2 equivalent emissions
		Billion gallons per year	Million acres	Million acres per bil. Gal per year	Grams CO2e per MJ of ethanol per year 1/
9 to 11 BG		2	.607	.304	13.6
11 to 13 BG		2	.655	.327	14.3
13 to 15 BG		2	.684	.342	14.5
2001 to 15 BG		13.23	4.258	.322	14.5
EU JRC-IE (Edwards et al., 2010) 6/					
	LEITAP			4.08	151.11
	AGLINK			2.41	89.32
	GTAP			.78	28.83 (62)
	IMPACT			.51	17.5

Na=not applicable.

1/ The carbon content of fuels, or the aggregate LCA emissions associated with production, including land-use change impacts, is often expressed relative to energy content so the carbon to energy ratio (gCO_2e/MJ) can be compared across fuels. A wide variety of methods and assumptions are used by the studies in converting land-use change impacts into associated estimates of carbon emissions; an explanation of such methods is beyond the scope of this report but critical to a full understanding of the carbon intensity estimates and why they differ among studies.

2/ Searchinger et al. reported their results in terms of a 55.92-billion-liter increase in ethanol production, which resulted in a 10.8-million-hectare change in global land use.

3/ Based on a 10-percent increase in U.S. ethanol use using 10-year averages of U.S. ethanol use and world crop area taken from the 2007 FAPRI baseline. Impact multiplier of 0.009 taken from Fabiosa et al. (2009), Table 2.

4/ Figures refer to international indirect land-use change only.

5/ Conversion to megajoules (MJ) of ethanol assumes each gallon of ethanol contains 76,330 Btu's of energy and each Btu is equal to 0.00105 MJ. Tyner et al (2010) derive results separately for multiple categories of scale of production, or billion gallons (BG) per year.

6/ The reported emissions figures associated with estimated land-use change are based on a central carbon stock estimate of 40 tC/ha of conversion and a 20-year carbon payback horizon. The JRC-IE report illustrates error bars around that estimate based on a range of 10-95 tC/ha. In the case of the GTAP model results, the JRC-IE calculated an additional emissions estimate based on a table of regional emissions factors that more finely reflects projections of different types of regional land conversion (with varying carbon stocks); the resulting emissions estimate is shown in parentheses.

A Chronology of Indirect Land-Use Change Estimates

In the February 2008 issue of *Science,* Searchinger et al. (2008a) published an influential early study of the effects of biofuel production on indirect GHG emissions. That study built on the Farrell et al. (2008) premise but used a more analytically rigorous estimation method to determine the land-use impacts of using corn for ethanol in the United States. The researchers used a worldwide agricultural trade model to explore how aggregate corn acreage diverted to ethanol feedstock production in the United States would translate into increased land use in the United States and elsewhere. Using a multi-market, multi-commodity international model of agricultural markets called the FAPRI (Food and Agricultural Policy Research Institute) model, Searchinger et al. assessed the land-use change and GHG implications of increasing corn ethanol production in the United States by 14.8 billion gallons. They projected that an additional 26.7 million acres of land would be brought into crop production worldwide (1.8 million acres per billion gallons of ethanol). The impact of domestic ethanol production was transmitted into global markets largely through price and export impacts on corn, wheat, and soybeans; prices were increased by 40 percent, 20 percent and 17 percent, respectively, while exports were estimated to decline by 62 percent, 31 percent, and 28 percent.[12]

As with all such estimates, Searchinger's (2008b) land-use change estimates are highly sensitive to the set of underlying assumptions used, including:

- Historical patterns of land conversion are used to estimate land conversion probabilities internationally (Searchinger used data collected at the Woods Hole Research Center estimating the proportion of newly converted cropland coming from different forest and grassland pools in different regions of the world).
- The analysis assumed that yields would continue to rise as they have in the past, but no additional price-induced yield increases were considered. (It was assumed that the impact of such increases would be cancelled out by greater use of lower-productivity marginal lands in production).
- The analysis employs a partial-equilibrium, one-world price model to generate projections on worldwide indirect land-use change. From their model, significant, market-driven, acreage responses emerge in China and India.

When the researchers change assumptions about land productivity and conversion efficiency, the estimated magnitude of land conversion and the resulting GHG pay-back period, are significantly reduced. The authors, however, also argue that their land conversion estimates may be low, due to an assumption that conversion of grassland has no further indirect impact (as grazing cattle are pushed into forest, for instance) (Searchinger, 2008b).

The increased concern about indirect land-use issues engendered by the initial studies by Farrell and O'Hare (2008) and Searchinger et al. (2008a) led researchers to use refined or adjusted estimation procedures to address some of the criticisms of that original analysis and provide additional perspective on the complexity of the issue. Gibbs et al. (2008) use a spatially explicit data set of crop locations and yields to explore in more detail the influence of changing crop yields and advances in conversion technology on biofuels-related land conversion. Although they demonstrate that land demand can be substantially reduced with yield and technology improvements, the authors echo Searchinger's finding that when biofuel production triggers conversion of tropical forests, the estimated payback times for GHG emissions remain on the order of 30-300 years.

In a subsequent analysis of the degree to which ethanol production in the United States would drive agricultural land-use change in other countries, Keeney and Hertel (2009) also focus on the sensitivity of yield-gain assumptions, but they add a consideration of bilateral trade patterns into their estimates of global supply response. This analysis uses a modified version of GTAP—a CGE model that considers production, consumption and trade of goods and services by region and globally across multiple sectors. Over the past several years, a team of researchers at Purdue University has refined and updated the GTAP model on an ongoing basis to support analysis of land-use change in response to biofuels policy. To capture the competition for land between land-use sectors, the GTAP model was augmented with a land-use module (GTAP-AEZ) that models the expansion of cropland and its distribution among different agricultural activities based on the price elasticity of yield and the ratio of productivities of marginal and average lands (Tyner et al., 2009). Other model modifications provided further refinement of the energy sector, including the three major biofuels (corn ethanol, sugarcane ethanol and biodiesel) and energy sector demand and supply elasticities that have been re-estimated and calibrated to 2006 data (Beckman and Hertel, 2009)

Keeney and Hertel (2009) conclude that assumptions about the responsiveness of U.S. yields to price are critical in determining the magnitude of acreage conversion in the rest of the world (ROW); when yields do not respond to price (as assumed by Searchinger), ROW acreage conversions are much higher. Using a range of yield elasticities that they describe as "plausible" based on past work and current agricultural conditions, they find that after 5 years, nearly 30 percent of the increased corn output can be met through yield increases rather than through acreage conversions with indirect repercussions. This study and several that followed have therefore concluded that yield-increasing technology plays a key role as a land substitute in analyses of the land demands of ethanol expansion. Assumptions about how/whether yields will continue to increase, and the role of biotechnology in boosting that increase, will strongly influence any estimation of future land conversion.

Keeney and Hertel (2009) also find that a consideration of bilateral trade patterns is critical in predicting patterns of acreage conversion. In a departure from the "one world price" philosophy of global response, they theorize that countries with a well-developed historical trading relationship with the U.S. are more likely to be affected, and to experience a market response, when prices and U.S. exports change (Hertel et al., 2010). Unlike the Searchinger results, which linked cropland conversion in the United States to acreage responses in Brazil, China, and India (as well as the United States), Keeney and Hertel's results project the most dramatic international acreage responses in Canada and Brazil. Because land-use policies and conversion patterns vary widely from country to country, location of acreage response can have a very important impact on associated environmental impacts of interest such as GHG emissions.

In a later study of biofuels-induced land allocation using FAPRI, Fabiosa et al. (2009) estimated that a 1 percent increase in U.S. ethanol use would result in a 0.009 percent increase in world crop area. Most of the increase in world crop area would come from an increase in world corn area as corn producers respond to projected drops in corn exports and an estimated 26 percent increase in world corn price. Brazil and South Africa would respond the most, with multipliers of 0.031 and 0.042, respectively, followed by Mexico, the United States, Thailand, and Egypt. More moderate acreage responses also would occur among other feed grains and soybeans. Based on the 10-year averages of U.S. ethanol use and world crop area taken from the 2007 FAPRI international baseline, and using the world area impact multiplier from Fabiosa et al. (2009) (0.009), the results suggest a land-use impact multiplier of 1.64 million acres per 1 billion gallons of additional ethanol use. This figure includes both the domestic and international cropland expansion expected as a result of increased ethanol production. The authors, however, add a caveat to their findings with the observation that data on the behavior of ethanol markets are limited, which makes it difficult to econometrically estimate the elasticities required by the biofuel market module. They also find their results are sensitive to the assumptions made about the ability of the livestock market to adapt to the use of DDGs in feed and the behavior of commodity stock adjustments in the short term.

Regulatory Estimation Efforts

The California Air Resources Board, in support of its recently adopted low carbon fuel standard, contracted with the Purdue research team to use the modified GTAP model described earlier to calculate how global patterns of land use would change globally in response to a 13.25 BGY increase in ethanol production, assuming a 2006 baseline for crop production patterns and conversion efficiency. CARB released its findings in March 2009; its results suggested each additional billion gallons of corn-starch-based ethanol would require only 726,000 acres; about 60 percent less than that suggested in Searchinger et al. (2008a).[13] Because completely different models are used to derive the results, it is difficult to directly attribute differences to specific assumptions, but Searchinger attributes some of the gap to varying assumptions about how world food demand and production would respond to increased prices (Charles, 2009). Searchinger's methodology assumed a modest response of food demand to world prices; CARB's finding of a more extreme food response buffers the land requirement impact of increased biofuel production. A more elastic food demand response, while keeping land demand low, may also exacerbate issues related to hunger and poverty (Charles, 2009).

In a parallel regulatory drive to quantify the GHG content of biofuels, the EPA integrated several models to explore the emissions from domestic and international land-use changes induced by increased renewable fuels consumption in the U.S. (See Figure 4.) For its analysis of the domestic response to biofuels production, EPA relied on the Forestry and Agriculture Sector Optimization Model (FASOM). FASOM estimated changes in domestic agricultural land use, as well as changes in domestic crop prices and crop export volumes. A parallel analysis used the integrated FAPRI models to project the responses of international agricultural markets and land use to the change in domestic activity. Because FAPRI does not address the locations

Measuring the Indirect Land-Use Change Associated ...

and types of land that comes into production within countries, country-specific estimates of conversion types were extrapolated from Winrock estimates of land-use conversion between 2001 and 2004 derived from satellite imagery (U.S.E.P.A., 2009).

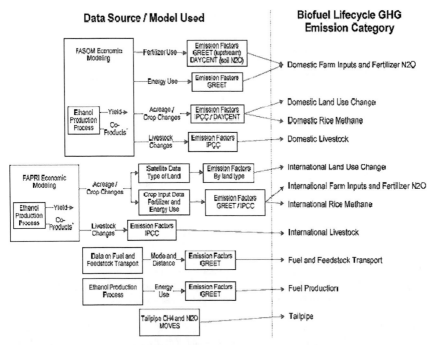

Figure 4. EPA system boundaries and models used. Source: Renewable Fuel Standard Program (RFS II) Regulatory Impact Analysis.

The EPA analysis produced two rounds of results for the domestic and international land-use changes associated with domestic biofuel production, corresponding to the proposed and final rules. The proposed rule was released in May 2009. The GHG emissions associated with international indirect land-use change were found to be a significant component of the overall GHG content of all of the biofuels analyzed; one set of results for corn ethanol is shown in Figure 5. According to these preliminary estimates, none of the corn ethanol production pathways satisfy the requirement that conventional corn ethanol reduce GHG emissions by 20 percent relative to gasoline to qualify under the Renewable Fuel Standard.

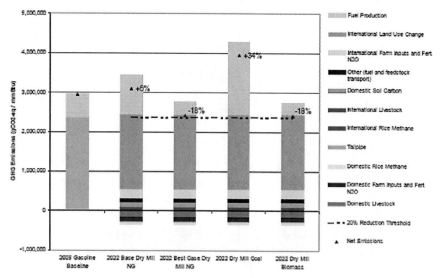

Figure 5. Corn ethanol life-cycle GHG results calculated for the 30-year, 0 percent discount rate scenario. Source: EPA Draft Regulatory Impact Analysis: Changes to Renewable Fuel Standard Program, http://www.epa.gov/otaq/renewablefuels/420d09001.pdf

Acknowledging the complexity and uncertainty inherent in deriving these numbers, EPA solicited public and expert feedback on the proposed rule throughout an extended 120-day public comment period and through use of an independent peer review process. The specific components of the methodology targeted for peer review were:

1) Land-use modeling, focusing on the use of satellite data to estimate land conversion probabilities and EPA's proposed land conversion GHG emission factors;
2) Methods to account for time in weighting GHG emissions and savings;
3) Methodology for calculating the GHG emissions from foreign crop production, including both the models and the data/assumptions used; and
4) The integration of the various models required to provide overall lifecycle GHG estimates.

A summary and compilation of the peer review results and the extensive public comments are available on the EPA website.[14] Table 2 gives a short description of the major changes between the proposed and final rules for RFS II.

In response to public and expert input during the public comment period, and to the availability of more recent data than were available during the proposed rule development, EPA made a number of changes to the assumptions and methodology underlying its life-cycle analysis (U.S.E.P.A., 2009). As a result, their estimates for international indirect land-use change associated with domestic biofuel production dropped from 1.692 acres per 1000 gallons of ethanol to .722 acres per 1000 gallons of ethanol.

The most significant of the EPA changes that affected the magnitude and location of projected indirect land-use demand, were the inclusion of induced corn yield increases in response to corn price increases, increased substitutability of distillers' grains with existing corn and soybean meals in beef cattle and dairy cow diets (which reduces demand and land requirements for other meals), improved spatial and temporal resolution of satellite data used for investigating land conversion internationally and more detailed modeling of Brazil's agricultural sector and land-use policies, and inclusion of forest and idle cropland as potential sources of domestic agricultural land. While the new assumptions regarding DDG prospects and yield projections had a substantial impact on land demand, some researchers argue that it is still unclear whether the new projected estimates represent improvements over the estimates used in the proposed rule (Plevin, 2010).

The reduction in the land-use change estimates associated with the revised RFSII assumptions had an impact on calculations used to determine whether corn ethanol satisfies the 20 percent reduction in GHG emissions required under the Renewable Fuel Standard. The final rule results for GHG emissions from ethanol production are shown in Figure 6. According to the revised estimates, in 2022 both corn ethanol production from natural-gas-powered plants and corn ethanol production from biomass-powered plants will satisfy the 20 percent reduction in GHG emissions required by the RFS. Note the significant reduction in the contribution of GHG emissions from international land-use change.

Table 2. Updates between proposed and final rules for RFS II

Updates to domestic agricultural sector modeling
- Incorporated results from the FASOM forestry module as well as the cropland module
- Added new land classifications: cropland, cropland-pasture, rangeland, forest-pasture, forest, Conservation Reserve Program (CRP), developed land
- Updated emissions factors for N2O and soil carbon
- Updated emissions factors for farm input production (fertilizer, etc.)

Updates to international agricultural sector modeling
- Incorporated a detailed Brazil module into the international model framework (including regional crop and pasture modeling)
- Added a factor representing price-induced yield changes
- Updated international agricultural GHG emission estimates
- Updated figures for Brazil's sugarcane production

Updates to biofuel processing in both domestic and international agricultural sector modeling
- Built in corn fractionation pathways (with coproduct markets, etc.)
- Adjusted DGS coproduct replacement rates to reflect more efficient use of DGS coproduct in livestock diets
- Added a coproduct credit for glycerin in biodiesel production
- Updated estimates of process energy use

Updates to land-use change modeling
- Used more recent and higher resolution satellite data with longer time coverage (2001-2007)
- Augmented satellite data with region-specific data where available (e.g., data from Brazil on pasture intensification)
- Used new soil carbon data
- Used new studies on long-term forest growth rates

Petroleum baseline updates
- Updated petroleum baseline to 2005

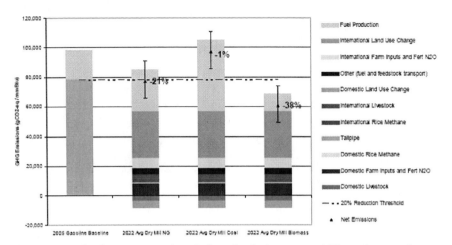

Figure 6. Results for new corn ethanol plants by fuel source and life-cycle stage for an average 2022 plant assuming 63 percent dry and 37 percent wet DDGs (with fractionation). Source: Renewable Fuel Standard (RFS II) Regulatory Impact Analysis (Final), http://www.epa.gov/otaq/renewablefuels/420r10006.pdf.

Although EPA and CARB came up with similar aggregate estimates of the indirect land-use requirements associated with domestic corn ethanol production (see Table 1), they used very different analytical structures and modeling frameworks to derive those estimates. EPA's analysis, for instance, measures the impact of biofuels policy relative to a "business as usual" case projected for the year 2022; parameter inputs, and estimated results, are assumed to reflect anticipated changes in crop yields, energy costs, and production plant efficiencies between now and 2022.[15] The CARB (2009) analysis, on the other hand, looks at land-use changes relative to a 2001 baseline year, which is the most recent year for which a complete global land-use database was available. The resulting estimates of land conversion are then corrected to account for the changes in agriculture observed to occur between 2001 and present. The most significant of those adjustments captures improved corn yields, which were observed to increase by 9.5 percent between 2001 and 2009. The CARB analysis therefore represents an estimate of the biofuel impacts on land demand assuming current crop production and conversion technologies as a baseline, while the EPA analysis generates impact estimates assuming a set of anticipated future technologies as a baseline.

In addition to overarching differences in the structure of the impact analysis and the way that baselines, or business-as-usual assumptions, are defined, the two regulatory efforts used completely different modeling frameworks to quantify the land-use changes associated with biofuel production. EPA used the FASOM/FAPRI combination to evaluate land use and GHG impacts, while CARB derived its estimates using the GTAP model. To explore the robustness of its land-use estimates to modeling framework, EPA performed a series of exploratory runs using a version of GTAP that it customized to mimic as closely as possible its FASOM/FAPRI scenarios. The results of that comparison (see Figure 7) highlight some of the differences in results generated by the two modeling frameworks.

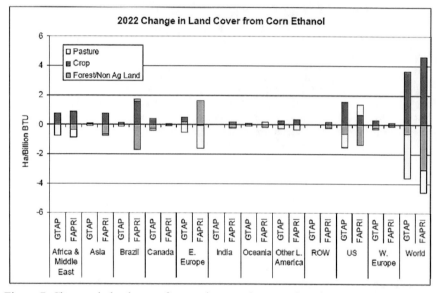

Figure 7. Changes in land cover from an increase in corn ethanol. Source: Renewable Fuel Standard (RFS II) Regulatory Impact Analysis (Final), http://www.epa.gov/otaq/renewablefuels/420r10006.pdf.

EPA's GTAP estimates suggest that, worldwide, a smaller amount of land will be converted to cropland per billion BTUs of ethanol and a greater proportion of that will come from pasture rather than forest, relative to a similar scenario derived using the FAPRI model. EPA identifies a few important factors that contribute to the difference in aggregate land demand (U.S.E.P.A., 2009):

- GTAP reflects a more optimistic assessment of the potential for intensification of agriculture as a result of higher prices, so that higher prices induced by renewable yield result in higher yields for both corn and other crops impacted by competition for land.

- GTAP is a general equilibrium model that explicitly models land demand in multiple sectors, so it captures the buffering impact on agricultural land demand that occurs when the prices of non-agricultural products rise and "push back" against the agricultural sector in its push to expand.
- The GTAP version used does not include a pool of "unmanaged" land that is available for cropland expansion. FAPRI allows land to come from a variety of sources such as grassland, savanna, shrubland, and wetlands, while GTAP assumes that all land that is not cropland or pasture is forest. If forestland is a relatively high-valued land use, that assumption could constrain the expansion of cropland acres.

The regional distribution of new cropland demand and land-cover change diverges between the two models as well. In particular, while FAPRI suggests there will be significant land-use change in Brazil, Asia, and Eastern Europe, GTAP does not. As mentioned earlier, GTAP was originally designed as a trade model, and its structure explicitly reflects historical trade relationships. GTAP results therefore have a tendency to maintain existing trade patterns and impose land-use demands on countries that have historically been major trading partners with the United States, while FAPRI results are more flexible with respect to how future trade patterns will respond to changes in global market conditions (U.S.E.P.A., 2009). Despite these differences, however, the EPA assessment of the modeling comparison concluded that "the GTAP model results were generally consistent with our FAPRI-CARD/satellite data analysis, in particular supporting the significant impact on international land use" (U.S.E.P.A., 2009).

Continuing Research

Spurred by the regulatory modeling efforts, research on this topic has continued to emerge and to evolve. Hertel et al. (2010) again used the modified GTAP model to explore four market-mediated responses to increased domestic ethanol production:

1) A reduction in global food consumption;

98 United States Department of Agriculture

2) An intensification of agricultural production, including increases in crop yields;
3) Land-use changes into cropping in the United States; and
4) Land conversion in the rest of the world.

To more explicitly illustrate the impacts of these critical factors, their analytical approach focused on determining the proportional influence of each of those factors on the resulting estimates of international land-use change. To isolate the land-use impacts of these individual factors, the study used an equilibrium model based on 2001 data, rather than a comparison against a forward projection of changes over time with multiple other variable forces operating on land uses simultaneously (the approach taken by EPA).

In exploring the sensitivity of their results to the factors just listed, Hertel et al. sequentially impose a series of assumptions regarding critical variables such as the yield elasticity (.25) and the relative productivity of new land brought into production (66 percent of the productivity of land currently in cropland), and calculated the impact of each additional factor on the land requirements associated with a 15 billion gallon per year increase in corn ethanol production. Figure8 illustrates the progressive effects on land demand of introducing market-mediated adjustment assumptions one at a time. The "gross" land requirement on the far left represents a case where no price or market responses are considered at all; producing corn for ethanol production requires an amount of land equal to the amount of corn required divided by the average yield. The researchers then sequentially added additional elements reflecting on particular aspects of market response and calculated the implications for land demand.

The order of the elements added into the analysis is as follows:

1) Price response arising from the introduction of constraints on the availability of suitable land, including a reduction in non-food demand and an intensification of livestock and forestry activities.
2) The ability of coproducts to substitute for feed meals and therefore reduce demand and land required for production or crops for those meals.
3) Response of world food demand to increase in prices (which is sensitive to assumptions made about price elasticity).
4) Inclusion of a yield elasticity that increases crop yields in response to price increases.

5) Assumption of a reduction in the productivity of newly converted land relative to existing cropland.

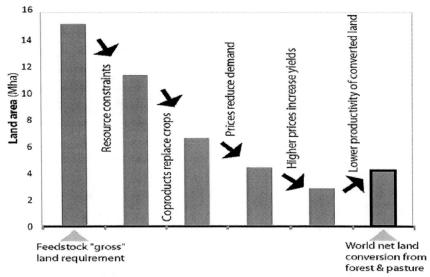

Figure 8. Estimates of indirect land-use change evolve as the scope of market-mediated impacts expands. Source: Hertel et al., 2010.

Because these factors interact in influencing land-use demand, the magnitude of impact for each factor is sensitive to the order in which market response elements are added into the analysis. Nevertheless, the figure effectively illustrates the way in which such factors contribute to the price responsiveness of supply and demand in the market and, taken together, determine the full market-mediated response to biofuels production.

When all market-mediated effects are considered, Hertel et al. (2010) found that despite a consideration of price-driven crop productivity improvements, domestic average yields of non-coarse grains and oilseeds drop as these crops are displaced from prime acreage and moved onto marginal acreage with reduced productivity. In the United States, they estimate that two-thirds of the new marginal acreage will come from forest cover, while one-third will come from pastureland. Internationally, the regions where production responds most vigorously to drops in corn exports are regions that are either significant importers of corn from the United States or that compete with U.S. corn exports in other markets, including Latin America, the EU, and China.

100 United States Department of Agriculture

While much of the research in this field reports out a single figure for average land-use impact, Hertel et al. (2010) analyze in more detail how the uncertainty associated with their parameter assumptions impacts their estimates of expansion and contraction of various land uses. In the supporting documentation for their research, the authors identify the following variables and parameters as the most critical ones in driving their estimates of land-use change:

1) The response of yields to price (yield elasticity);
2) The ease with which land moves between alternative uses, including cropland, pasture, and forestry (elasticity of land transformation across uses);
3) The assumed yield from marginal lands (elasticity of effective cropland with respect to harvested cropland);
4) The response of other countries' land markets to shocks in U.S. exports and prices (trade elasticities for crops and other food products); and
5) The extent to which imports can substitute for one another when one commodity or product experiences a shock (elasticity of substitution among imports from different sectors).

Their analysis suggests that the most uncertain results are the forestry conversion estimates, particularly in areas outside of the U.S., Canada, and the EU.

Hertel et al. (2010) also vary their parametric assumptions to compute bounding values for the emissions impacts associated with their land-use change analysis. The resulting lower and upper bounds, 15 and 90 gCO2 per MJ per year, suggest that the distribution of possible land-use change values is not symmetrically distributed around the average value of 27. The tail of the distribution of possible emissions values is much longer on the side of increased emissions; in other words, if actual emissions deviate from the estimated average on the low side, they may be somewhat lower than the projection, but if they deviate on the high side, they may be significantly higher than the projection. The authors therefore conclude that, "Better understanding of skewness and long tails in an estimated distribution of the ILUC value will probably imply that an optimal value for the index assigned to a particular biofuel *will be different from* a central estimator of its ILUC effect."

Measuring the Indirect Land-Use Change Associated ...   101

The most recent estimates of biofuels-related land-use change to emerge from the Purdue research team were released in April 2010 (with revised estimates released in July 2010). Tyner et al. (2010) compare the land-use implications under three different sets of parameter assumptions:

1) An analysis of impacts relative to the 2001 database;
2) An analysis of impacts relative to a 2006 database that is calculated as having evolved from the 2001 database in a world with regional changes to GDP, gross capital formation, population, biofuel production, crop yields, and forest area; and
3) An analysis of impacts relative to the 2006 baseline but with continued growth in crop yield and population for ethanol amounts corresponding to production years beyond 2006.

The land conversion requirements drop sequentially as the authors move through the scenarios, largely due to the impacts of changes in assumptions about crop yields. Scenario 2 reflects observed increases in crop yields over the period 2001-2006, while scenario 3 further assumes an annual 1 percent increase in crop yields for all crops in all regions beyond 2006. The reduced land requirements in scenario 3 reflect a set of underlying parameter assumptions that result in land-saving yield impacts dominating the increased land demand associated with increased food demand and a growing population. These results, of course, are sensitive to the underlying assumptions; there may be some debate, for instance, about whether it is reasonable to expect a 1 percent annual increase in crop yields for all crops in all regions.

The land-use change requirements results for scenario 3 are shown in Table 1; 1000 gallons of ethanol production requires an average of .32 acres of additional land. The land requirements for biofuel production are significantly lower than those found in other studies, including those performed using GTAP. The authors describe two recent modifications to the model that contribute to this reduction (Tyner et al., 2010). The first is the introduction of two new land categories into GTAP's available land pools: cropland pasture and unused cropland, including acreage enrolled in the CRP. Land brought into production from these land pools is not considered "land conversion," so using these lands reduces the amount of forest or grassland that must be converted. The second modification increased the refinement with which the productivity of marginal lands is considered in the model. In prior GTAP analyses, newly converted land is generally assumed to have 66 percent of the

102   United States Department of Agriculture

productivity of existing cropland. The revised model, however, uses a process-based biophysical simulation model to calculate a set of regional productivity factors at the agro-ecological zone (AEZ) level, some of which are larger than 66 percent. Over the three scenarios, the authors find that the fraction of land-use change that occurs in the United States varies between 24 and 34 percent, while the percentage of new cropland that comes from forest ranges between 25 and 33 percent.

In the summer of 2010, the European Commission's Joint Research Centre published a series of consultation documents that had been contracted to inform decisions on how indirect land use effects should be handled in implementation of the EU Renewable Energy and Fuel Quality Directives (RED/FQD). One of those documents—"Indirect Land Use Change from increased biofuels demand: Comparison of models and results for marginal biofuels production from different feedstocks"—describes the extensive modeling effort conducted, some scenarios of which looked at the impacts of U.S. grain-based ethanol production (Edwards et al., 2010). This effort employed a range of models; some of the results of the EU analysis for U.S.-based ethanol production from maize or coarse grains are shown in Table 1.

An indepth description of the EU modeling effort, and an explanation of the results, can be found in the source document listed above. However, the report highlights the following reasons for the discrepancies between model results:

- The IFRPI-IMPACT model assumes a large price-induced yield gain across crops, which results in relatively low area changes.
- GTAP assumes significant contributions from price-induced yield gains as well, but that effect is countered by the fact that GTAP assumes lower yields for crops produced on newly converted production areas.
- Three of the models employ some type of Armington trade assumptions (GTAP, LEITAP and AGLINK) to introduce "stickiness" into the global trade response to represent transport costs, import tariffs and regulations, and information flows. This stickiness affects the extent to which crop production can be shifted to developing countries, in comparison to the integrated world market assumed by IMPACT.
- LEITAP does not account for byproduct effects, which increases land-use impacts relative to the other models.

- Significant differences exist among the models in how they calculate the area change required for an increase in crop production. Only GTAP, for instance, calculates the incremental area required based on an assumed marginal/average yield ratio (.66) that accounts for the fact that crop areas are expanding onto marginal lands with the potential for lower yields.
- Significant differences exist in the extent to which food demand declines as prices rise. LEITAP assumes very little food demand response, while the land-use change impacts predicted by both IMPACT and GTAP drop by more than 50% when their food demand response assumptions are implemented.

The EU modeling effort illustrates the spectrum of potential sources of variability and uncertainty across models and parameter assumptions. While the land-use impact estimates derived reflect this variability, all of the estimates across models, feedstocks, and type of biofuel, show significant increases in land-use requirements for crops resulting from scaled-up biofuel demand (Edwards et al., 2010).

VI. SUMMARY AND FUTURE RESEARCH NEEDS

The increase of biofuel production to reduce energy dependence on fossil fuels led to concerns about unintended consequences of that activity. In particular, life-cycle analyses that accounted for greenhouse gas emissions identified the importance of considering the underlying land-use changes that might be associated with increased feedstock production. Estimating these changes, however, has been a daunting task. Mathematical models are used to indicate numerically how the aggregate behavior will change due to new conditions or policies, and they serve as a proxy for what cannot be directly observed. However, no single model currently exists that can address all of the questions related to this issue.

Uncertainty is an unavoidable aspect of policy impact modeling that makes projections into the future. There are many drivers of land-use change in the U.S. and internationally that exist regardless of biofuel policy. The assumptions embedded within the different models estimating land-use change will affect model results. In the past few years, the analytic approaches used in these studies have been expanded and the underlying assumptions have been

104 United States Department of Agriculture

made more transparent. Many of the differences in model estimates can be traced to differences in assumptions about:

- Crop yields and the projected elasticity of response to demand-driven price increases;
- The baseline that is used from which to measure change;
- The anticipated productivity of newly converted land and the amount of land required to meet increased production demands by region;
- The structure and flexibility of trade flows;
- The price elasticity of demand for agricultural food and feed products;
- The scope of the life-cycle assessment–including, for instance, whether the livestock and forest sectors are explicitly modeled as competitors for land.

Managing uncertainty in the context of modeling indirect land-use change has involved:

- Identifying the variables that are particularly important in contributing to the uncertainty of estimates and improving the precision with which such variables are represented with the analysis (e.g., future crop yields, the productivity of newly converted lands, and the substitutability of DDGs in livestock diets);
- Identifying relevant relationships that require more refined analysis (such as the importance of trade relations in determining likely sources of increased agricultural production);
- Understanding the nature of the remaining uncertainty, its effects on the distribution of potential outcomes, and the implications of incorporating different measures of that uncertainty into policy;
- Designing policy to ensure that existing regulations evolve as the science becomes more sophisticated.

Both EPA and CARB call for regular examination and updating of the indirect land-use impacts component of the GHG quantification methodology. EPA includes in its final rule the following stipulation: "EPA will request that the National Academy of Sciences evaluate the approach taken in this rule, the underlying science of life-cycle assessment, and in particular indirect land-use change, and make recommendations for subsequent lifecycle GHG assessments on this subject" (EPA, 2010). The CARB resolution also states, "... that the Board directs the Executive Officer to convene an expert workgroup to assist the Board in refining and improving the land use and indirect effect analysis of transportation fuels and return to the Board no later than January 1, 2011 with regulatory amendments or recommendations, if appropriate, on approaches to address issues identified" (CARB, 2009).

Establishing such mechanisms for continuous refinement is a critical element of policy design in this area. The science underpinning these regulations is an evolving discipline, and the estimates derived from it will continue to improve, as will the policy mechanisms established to handle ongoing uncertainty. Additional research on variables that have been identified as critical drivers in determining the land-use impacts of biofuels policy also will be instrumental in refining modeling capacity and estimates in this area. Research and data that can facilitate greater understanding of critical dynamics in land-use impact modeling includes:

- Improved data on land use and land cover change worldwide;
- Improved data on the extent and productivity of existing and potential cropland worldwide, including previously cleared, "degraded," or underutilized lands;
- Greater understanding of and ability to model prospective growth in crop demand and supply by region worldwide. This includes refined analysis of development and adoption of crop productivity technologies such as biotechnology as well as impacts of income, population growth, and dietary transitions;
- Improved modeling and parameter estimates around the substitutability of biofuel coproducts in other markets, such as livestock feed markets.

Future research that addresses more explicitly the sources and magnitude of this uncertainty will be particularly useful in informing policy design. Hertel et al. (2010) set a precedent for this by highlighting the need for performing and reporting more extensive sensitivity analysis in research on land-use impacts. These authors state that a more explicit treatment of sensitivity and distributions of possible outcomes is also needed to inform a more sophisticated treatment of uncertainty within policy design itself. They suggest that central estimators such as average or median values may not be the most appropriate values to assign to land-use change or emissions estimates in cases where highly asymmetrical distributions of possible outcomes exist (Hertel et al. 2010). Further research on alternative estimators or alternative methods of capturing distributions within a policy indicator,

106 United States Department of Agriculture

together with research on how those methods correspond to particular attitudes toward risk and risk management, may lead to a critical development in the appropriate incorporation of inherently uncertain estimates into policy such as that related to biofuels and land-use change.

As the science of projecting land-use change evolves, USDA has an important role to play both in supporting ongoing EPA efforts and in developing additional research capacity for exploring the critical variables determining the direct and indirect land-use impacts of domestic agricultural production and the way they play out through domestic and global market interactions. Existing ERS research corroborates the upwards pressure that biofuel production puts on commodity prices and land demand (Malcolm et al. 2009) and the sensitivity of domestic bioenergy's economic and welfare impacts to uncertain projected responses within world energy markets (Gehlar et al. 2010).

In an ongoing effort to integrate more sophisticated land allocation considerations into its analyses of bioenergy production, USDA's Economic Research Service is expanding two in-house agricultural models to more explicitly represent the critical variables driving the relationship between land use and biofuels and to differentiate among potential pools of land for production. The Regional Environment and Agriculture Programming Model, a detailed domestic partial equilibrium model of cropland and livestock agriculture, is being expanded to include a forestry component and a more sophisticated treatment of competition for land and conversion among uses. The Future Agricultural Resources Model, a global general equilibrium model that was a pioneer in the movement to partition land as an economic input into different land classes, is being retooled to accommodate a dynamic forestry sector, to explicitly track energy technologies and energy accounting, and to allow for cellulosic conversion technologies that create biofuels from feedstocks such as crop residue, switchgrass, or fast-growing trees.

ERS models without explicit land-use analysis capacity are also being modified to support research related to trade and market response to biofuels production that can then be used to inform future land-use analyses. The U.S. Applied General Equilibrium (USAGE) Model includes more than 20 major importing and exporting trade partners and the data are being updated to capture the supply response of the farm sector to address bioenergy-related issues. The Partial Equilibrium Agricultural Trade Simulator (PEATSim) is being enhanced to include more domestic and international policies affecting major crop and oilseed markets, as well as oilseed product markets, sugar, livestock, and dairy.

Continued research and modeling efforts will be required to narrow the bands of uncertainty associated with projections of land-use change and domestic policy. New models, model refinements, and improved data will all help increase the precision with which input parameters are estimated and behavioral relationships are represented. Still, the successful integration of science and policy must come with the recognition that future projections will always carry some degree of uncertainty suggesting that policy design must accommodate uncertainty. As research moves forward, an explicit focus on the nature and structure of input and output uncertainty, and on the full distributions of possible outcomes and estimates that result, will facilitate improvements to policy design over time.

REFERENCES

Alfredsson, E.C. 2004. "'Green' Consumption – No Solution for Climate Change," *Energy*, 29(4); 513-524.

Angelsen, Arild and David Kaimowitz. 1999. "Rethinking the Causes of Deforestation: Lessons from Economic Models," *World Bank Research Observer*, 14(1), February.

Armington, Paul. 1969. "A Theory of Demand for Products Distinguished by Place of Production," Staff Papers - International Monetary Fund, 16(1), March.

Beckman, Jayson, and Thomas Hertel. 2009. "Why Previous Estimates of the Cost of Climate Mitigation Are Likely Too Low," Purdue University, GTAP Working Paper No. 54.

Beckman, Jayson, Carol Jones, and Ron Sands. 2011 Forthcoming. "A Global General Equilibrium Analysis of Biofuel Mandates and Greenhouse Gas Emissions," *American Journal of Agricultural Economics*, January.

Biomass Research and Development Board. 2008. *Increasing Feedstock Production for Biofuels: Economic Drivers, Environmental Implications, and the Role of Research*, Report submitted by the Interagency Feedstocks Team, December.

Brander, M., R. Tipper, C. Hutchison, and G. Davis. 2008. "Consequential and Attributional Approaches to LCA: A Guide to Policy Makers with Specific Reference to Greenhouse Gas LCA of Biofuels," Technical Paper, Econometrica Press.

California Air Resources Board. 2009. "Proposed Regulation to Implement the Low Carbon Fuel Standard, Vol 1." Staff Report: Initial Statement of Reasons. 374 pp. Available online at http://www.arb.ca.gov/regact/2009/lcfs09/lcfsisor1.pdf.

Charles, Dan. 2009. "Corn-Based Ethanol Flunks Key Test." *Science* 324: p. 587.

Edwards, Robert, Declan Mulligan, and Luisa Marelli. 2010. "Indirect Land Use Change from Increased Biofuels Demand: Comparison of models and results for marginal biofuels production from different feedstocks." European Commission Joint Research Centre Scientific and Technical Report, Institute for Energy (JRC-IE). Available online at http://ec.europa.eu/energy ing_comparison.pdf.

Fabiosa, Jacinto F., John C. Beghin, Fengxia Dong, Amani Elobeid, Simla Tokgoz, and Tun-Hsiang Yu. 2009. "Land Allocation Effects of the Global Ethanol Surge: Predictions from the International FAPRI Model," Iowa State University, CARD Working Paper 09-WP 488, March.

Fargione, Joseph, Jason Hill, David Tilman, Stephen Polasky, and Peter Hawthorne. 2008. "Land Clearing and the Biofuel Carbon Debt," *Science*, 319: 1235-1238, 29 February.

Farrell, Alex, and Michael O'Hare. 2008. "Greenhouse gas (GHG) emissions from indirect land use change (ILUC)." Memorandum to John Courtis, California Air Resources Board. Available online at http://www.arb.ca.gov/fuels/lcfs/011608ucb_luc.pdf.

Fisher, Anthony C., and W. Michael Hanemann. 1990. "Option Value: Theory and Measurement," *European Review of Agricultural Economics*, 17: 167-180.

Fuglie, Keith O. 2010. "Accelerated Productivity Growth Offsets Decline in Resource Expansion in Global Agriculture," *Amber Waves*, Vol. 8, Issue 3, U.S. Department of Agriculture, Economic Research Service, September.

Gehlar, Mark, Agapi Somwaru, Peter Dixon, Maureen Rimmer, and Ashley Winston. 2010. "Economywide Implications from U.S. Bioenergy Expansion," American *Economic Review: Papers & Proceedings*, 100: 172-177.

Gibbs, Holly, Matt Johnston, Jonathon Foley, Tracey Holloway, Chad Monfreda, Navin Ramankutty and David Zaks. 2008. "Carbon Payback Times for Crop-Based Biofuel Expansion in the Tropics: the Effects of Changing Yield and Technology," *Environmental Research Letters*. 3(3), July.

Golub, Alla, Thomas W. Hertel, and Brent Sohngen. 2009. "Land Use Modeling in Recursively-Dynamic GTAP Framework," Chapter 10 in *Economic Analysis of Land Use in Global Climate Change Policy*, edited by Thomas W. Hertel, Steven Rose, and Richard S.J. Tol, Rouledge Press.

Gurgel, Angelo, John Reilly, and Sergey Paltsev. 2007. "Potential Land Use Implications of a Global Biofuels Industry," *Journal of Agricultural & Food Industrial Organization*, Vol 5.

Heisey, Paul W. 2009. "Science, Technology, and Prospects for Growth in U.S. Corn Yields," *Amber Waves*, Vol. 7, Issue 4, U.S. Department of Agriculture, Economic Research Service, December.

Hertel, Thomas W., Alla A. Golub, Andrew D. Jones, Michael O'Hare, Richard J. Plevin, and Daniel M. Kammen. 2010. "Effects of US Maize Ethanol on Global Land Use and Greenhouse Gas Emissions: Estimating Market-mediated Responses," *BioScience* 60(3): 223-231, March.

Hertel, Thomas W., Wallace E. Tyner, and Dileep K. Birur. 2010. "The Global Impacts of Biofuels Mandates," *The Energy Journal*, 31(1): 75-100.

Hertel, Thomas W., Steven Rose and Richard S.J. Tol. 2009. "Land Use in Computable General Equilibrium Models: An Overview." Chapter 1 in *Economic Analysis of Land Use in Global Climate Change Policy*, edited by Thomas W. Hertel, Steven Rose, and Richard S.J. Tol, Rouledge Press.

Isik, M. and W. Yang. 2004. "An Analysis of the Effects of Uncertainty and Irreversibility on Farmer Participation in the Conservation Reserve Program," *Journal of Agricultural and Resource Economics*, 29 (2): 242-259.

Keeney, Roman, and Thomas E. Hertel. 2009. "The Indirect Land Use Impacts of United States Biofuel Policies: The Importance of Acreage, Yield, and Bilateral Trade Responses," *American Journal of Agricultural Economics*, 91(4): 895-909, November.

Leibtag, Ephraim. 2008. "Corn Prices Near Record High, But What About Food Costs," *Amber Waves*, Vol. 6, Issue 1, U.S. Department of Agriculture, Economic Research Service, February.

Lubowski, Ruben N., Marlow Vesterby, Shawn Bucholtz, Alba Baez, and Michael J. Roberts. 2006. *Major Uses of Land in the United States, 2002.* Economic Information Bulletin No. 14, U.S. Department of Agriculture, Economic Research Service, May.

Malcolm, Scott A., Marcel Aillery, and Marca Weinberg. 2009. *Ethanol and a Changing Agricultural Landscape*, Economic Research Report No. 86, U.S. Department of Agriculture, Economic Research Service, November.

Marshall, Liz. 2009. "Biofuels and the Time Value of Carbon: Recommendation for GHG Accounting Protocols" WRI Working Paper. World Resources Institute, Washington DC. April.

Melillo, Jerry M., Angelo C. Gurgel, David W. Kicklighter, John M. Reilly, Timothy W. Cronin, Benjamin S. Felzer, Sergey Paltsev, C. Adam Schlosser, Andrei P. Sololov, and X. Wang. 2009. *Unintended Environmental Consequences of a Global Biofuels Program*, MIT Joint Program on the Science and Policy of Global Change Report No. 168, January.

Plevin, Richard. 2010. "Review of final RFS2 Analysis." Comments available online at http://plevin.berkeley.edu/docs/Plevin-Comments-on-final-RFS2-v7.pdf.

Roberts, Michael J. and Wolfram Schlenker, 2010. "The U.S. Biofuel Mandate and World Food Prices: An Econometric Analysis of the Demand and Supply of Calories." Presented at the National Bureau of Economic Research conference "Agricultural Economics and Biofuels." Available online at http://www.nber.org/confer/2010/AGs10/summary.html.

Searchinger, Timothy, Ralph Heimlich, R.A. Houghton, Fengxia Dong, Amani Elobeid, Jacinto Fabiosa, Simla Tokgoz, Dermot Hayes, and Tun-Hsiang Yu. 2008a. "Use of US Croplands for Biofuels Increases Greenhouse Gases Through Emissions from Land-Use Change," *Science* 319: 1238-1240.

Searchinger, Tim. 2008b. "Evaluating Biofuels: The Consequences of Using Land to Make Fuel," The German Marshall Fund of the United States Policy Brief. Available online at
http://www.gmfus.org/galleries/ct_publication_attachments/Econ_Searchinger_Biofruel. pdf.

Schmidt, Jannick H. 2008. "System delimitation in agricultural consequential LCA," *International Journal of Life Cycle Assessment* 13: 350-364.

Song, F., J. Zhao, and S. Swinton. 2009. "Switching to Perennial Crops Under Uncertainty and Costly Reversibility," Michigan State University, Department of Agricultural, food, and Resource Economics, Staff Paper No. 2009-14, Devember.

Stavins, R.N. 1999. "The Costs of Carbon Sequestration: A Recealed Preference Approach," *American Economic Review*, 89 (4): 994-1009.

Trostle, Ronald. 2008. *Global Agricultural Supply and Demand: Factors Contributing to the Recent Increase in Food Commodity Prices*, Outlook Report WRS-0801, U.S. Department of Agriculture, Economic Research Service, July.

Tyner, Wallace E., Farzad Taheripour, and Uris Baldos. 2009. "Land Use Change Carbon Emissions Due to US Ethanol Production," Report to Argonne National Laboratory, January.

Tyner, Wallace E., Farzad Taheripour, Qianlai Zhuang, Dileep Birur, and Uris Baldos. 2010. "Land Use Changes and Consequent CO_2 Emissions Due to US Corn Ethanol Production: A Comprehensive Analysis" Department of Agricultural Economics, Purdue University. Final Report (revised) to Argonne National Laboratory. Available at:
http://www.transportation

U.S. Department of Agriculture, Office of Chief Economist and World Agricultural Outlook Board. 2010. *USDA Agricultural Projections to 2019*, Prepared by the Interagency Agricultural Projections Committee.

U.S. Department of Energy, Energy Information Administration. 2010. *Annual Energy Outlook*. May.

U.S. Environmental Protection Agency. 2009. EPA Draft Regulatory Impact Analysis: Changes to Renewable Fuel Standard Program. 822 pp. http://www.epa.gov/otaq/renewablefuels/420d09001.pdf .

U.S. Environmental Protection Agency. 2010. EPA Final Regulatory Impact Analysis: Renewable Fuels Standard Program (RFS2). 1120 pp. Available online at
http://www.epa.gov/otaq/renewablefuels/420r10006.pdf.

Zilberman, David, Gal Hochman, and Deepak Rajagopal. 2010. "Indirect Land Use: One Consideration Too Many in Biofuel Regulation," *Agricultural and Resource Economics Update*, Giannini Foundation of Agricultural Economics, University of California, 13(4): 1-4, March/April.

GLOSSARY OF ACRONYMS

AEO	Annual Energy Outlook
AEZ	Agro-Ecological Zone
CARB	California Air Resources Board
CGE	Computable General Equilibrium
CO_2	Carbon Dioxide
CRP	Conservation Reserve Program
DDGs	Distillers' Dried Grains
DOE	U.S. Department of Energy
EIAEnergy	Information Administration
EISA	Energy Independence and Security Act

EPA	U.S. Environmental Protection Agency
EPACT	Energy Policy Act of 2005
ERS	USDA Economic Research Service
FAIR	Federal Agriculture Improvement and Reform
FARM	Future Agricultural Resources Model
FAPRI	Food and Agricultural Policy Research Institute
FASOM	Forestry and Agricultural Sector Optimization Model
GHG	Greenhouse Gas
GTAP	Global Trade Analysis Project
ILUC	Indirect Land Use Change
IMAGE	Integrated Model to Assess the Global Environment
LCA	Life-Cycle Analysis
LCFS	Low Carbon Fuel Standard
MJ	Megajoule
PE	Partial Equilibrium
PEATSIM	Partial Equilibrium Agricultural Trade Simulator
REAP	Regional Environmental and Agricultural Production
RIA	Regulatory Impact Analysis
RFS	Renewable Fuel Standard
RFS II	Renewable Fuel Standard - 2009
ROW	Rest of World
RTFO	Renewable Transport Fuel Obligation
USAGE	U.S. Applied General Equilibrium
USDA	U.S. Department of Agriculture

End Notes

[1] EISA section 201 amends section 211(o)(1) of the Clean Air Act to provide a new definition of "lifecycle greenhouse gas emissions" that includes "direct emissions and significant indirect emissions such as significant emissions from land use changes.

[2] Full text: http://thomas.loc.gov/cgi-bin/cpquery/T?&report=hr181&dbname=111&). Passed June 18, 2009.

[3] For a more thorough discussion of increased biofuel production, see the Biomass Research and Development Board (2008) report, *Increasing Feedstock Production for Biofuels: Economic Drivers, Environmental Implications, and the Role of Research.* USDA's OCE and ERS (along with other USDA Agencies, EPA, and DOE) participated in this report.

[4] Section 1504 of the Energy Policy Act of 2005 eliminated the requirement that reformulated gasoline (RFG) contain minimum levels of oxygenates. Current air quality requirements can usually be met without the use of oxygenates, and the associated demand for RFG oxygenates is much smaller than it was prior to the EPAct.

[5] Ethanol industry advocates have argued that regulatory limitations on the amount of ethanol that can be blended into gasoline effectively limits the amount of ethanol that can be sold in low-level ethanol blends to 12 billion gallons per year (or roughly 10% of U.S. gasoline consumption by volume). This "blend wall" for low-level ethanol blends may create an obstacle to absorption of the regulated levels of ethanol by the market. On October 13, 2010, EPA issued a partial waiver to allow fuel and fuel additive manufacturers to introduce gasoline that contains greater than 10 volume percent (vol%) ethanol and up to 15 vol% ethanol (E15) for use in certain motor vehicles. However, extensive market penetration of E15 will require changes to state laws, recommendations from vehicle manufacturers and adoption by fuel distributors.

[6] Most ethanol processing plants currently combust coal or natural gas to generate power. Cellulosic biorefineries can generate their own power by burning non-fermentable lignin byproducts from their biomass feedstock, considerably reducing their dependence on fossil fuels.

[7] For a summary of the energy-equivalent reductions of different feedstocks compared with fossil fuels, see Biomass Research and Development Board, 2008, pp. 81-83.

[8] Economists define elasticity as the responsiveness of the quantity demanded (supplied) of a good or service to a change in its price. More precisely, it gives the percentage change in quantity demanded (supplied) in response to a 1-percent change in price (holding constant all the other determinants of demand (supply). When demand (supply) is elastic (greater than one), demand (supply) is very sensitive to price. The fewer substitutes that exist for a good, the lower the price elasticity will be (the less responsive demand will be to a price change). Time is also a consideration in determining both consumer and producer price responsiveness for many items. The longer people have to make adjustments, the more adjustments they will make.

[9] There may be regions where newly converted cropland is as productive as existing cropland (e.g., parts of Brazil), but those cases may be the exception. Furthermore, converted areas may be more productive in the short run than in the long run. Estimates of land-use demand will greatly benefit from improved data on the productivity of converted land for cropland worldwide.

[10] Option value is the gain from being able to learn about future benefits that would be precluded by the conversion of land to an irreversible or partially irreversible use–the gain from retaining the option to continue with the current use and/or change uses in the future (Fisher and Hanemann, 1990).

[11] Assuming that the non-land-use-related carbon savings from corn ethanol is a 20 percent reduction from gasoline (19 gCO2/MJ), this conversion cost would require approximately 866 years of corn ethanol production to pay back. This number, though spectacular, is meant only to illustrate the rough magnitude of potential impact, as it is based on a set of unlikely "worst case" assumptions.

[12] Searchinger et al. (2008a) calculated that the land-use change associated with corn-based ethanol has a carbon impact in the range of 103 gCO2eq/MJ. These emissions alone were larger than the estimated emissions from gasoline (92 gCO2eq/MJ). In contrast, when emissions from land-use change were not included in the estimate of GHG content, corn-starch based ethanol reduced GHG emissions by 20 percent compared to gasoline.

[13] Primarily as a result of this reduced acreage, CARB estimated the GHG emissions associated with land-use change were 70 percent less than those estimated by Searchinger et al. The GHG emissions due to land-use change were reduced from 104 grams of CO_2 equivalent per MJ of ethanol to 30 grams of CO2 equivalent per MJ of ethanol.

114   United States Department of Agriculture

[14] http://www.epa.gov/otaq/renewablefuels/420r10003.pdf

[15] The final year in which the RFS II renewable fuel volume mandates are phased in is 2022.

INDEX

A

access, 38, 80
accounting, 12, 54, 56, 63, 68, 106
additives, 35, 58
agencies, 41
aggregate demand, 63
agricultural market, 58, 87, 90
agricultural producers, 27, 39
agricultural sector, 20, 54, 67, 68, 78, 93, 94, 97
agriculture, 16, 18, 20, 21, 36, 53, 55, 66, 68, 74, 75, 78, 80, 95, 96, 106
air pollutants, 15
air quality, 15, 29, 112
alternative energy, 81
analytical framework, viii, 25, 26, 53, 56, 68
apples, 57
appropriations, 55
aquifers, 20
Asia, 97
assessment, 27, 30, 61, 65, 96, 97, 104
atmosphere, 13, 14, 61
authority, 37

B

bacteria, 14
barriers, 25, 39, 79
base, vii, 3, 6, 11, 54

beef, 63, 93
behaviors, 5, 8, 34, 37
benefits, 4, 5, 11, 14, 15, 25, 28, 29, 32, 33, 35, 36, 60, 64, 65, 66, 75, 113
benzene, 58
bilateral relationship, 79
biodiesel, 56, 59, 88, 94
biodiversity, viii, 3, 4, 12, 22, 23, 25, 29, 52, 56
bioenergy, 4, 8, 10, 12, 15, 16, 20, 21, 22, 23, 24, 25, 26, 27, 40, 57, 73, 74, 75, 81, 82, 106
biological processes, 12
biomass, vii, 3, 4, 5, 7, 10, 11, 12, 15, 17, 18, 19, 20, 21, 22, 23, 26, 28, 33, 34, 35, 36, 37, 39, 40, 41, 58, 59, 60, 61, 75, 82, 93, 113
biotechnology, 8, 74, 89, 105
biotic, 16, 17
blend wall, 113
blends, 10, 113
bounds, 100
Brazil, 27, 42, 63, 81, 89, 93, 94, 97, 113
breakdown, 58
breeding, 74
buyers, 79
by-products, 8

Index

C

candidates, 30
carbohydrate, 57
carbon, 4, 11, 13, 14, 15, 16, 17, 18, 25, 29, 30, 31, 33, 55, 60, 74, 80, 82, 85, 86, 90, 94, 113
carbon dioxide, 13, 55
carbon emissions, 80, 85
carbon monoxide, 15
catchments, 21
cattle, 88, 93
causal relationship, 61
cellulose, 58, 59, 60
cellulosic biofuel, vii, 3, 5, 7, 8, 10, 11, 12, 15, 18, 23, 26, 27, 28, 30, 34, 40, 41, 58, 60, 61
Chad, 108
challenges, 8, 10, 25, 38, 39, 41, 77
changing environment, 7
chemical, 15, 17
chemicals, 65
China, 87, 89, 99
classes, 106
Clean Air Act, 57, 112
climate, 2, 4, 8, 13, 17, 18, 19, 21, 24, 26, 28, 31, 33, 36, 54, 58
climate change, 4, 8, 13, 17, 21, 24, 26, 28, 31, 33, 54, 58
CO2, 13, 14, 15, 61, 68, 84, 85, 111, 113
coal, 81, 113
cogeneration, 31
colonization, 67
combustion, 15, 57, 58, 61
commercial, 20, 28, 60, 66, 75
commercial crop, 75
commodity, 27, 52, 53, 64, 65, 66, 68, 71, 78, 79, 83, 87, 90, 100, 106
commodity markets, 28, 66
communication, 6, 38
communities, 2, 5, 10, 11, 25, 32, 34, 35, 36, 37, 38, 39, 40, 41
community, 8, 34, 35, 36, 37, 38, 39, 40
community support, 39, 40

competition, viii, 4, 8, 19, 29, 35, 36, 37, 52, 55, 61, 64, 66, 68, 73, 78, 81, 88, 97, 106
competitive advantage, 64
competitiveness, 26, 27
competitors, 104
compilation, 93
complex interactions, 4, 31, 66
complexity, 5, 30, 36, 71, 88, 92
compliance, 34
composition, 77
computation, 68
conference, 110
conflict, 81
consensus, 38, 78
conservation, 16, 19, 20, 23, 31, 35, 37, 78
construction, 10, 18, 30, 67
consumer taste, 64
consumers, 27, 41, 54, 78, 79
consumption, 17, 28, 30, 78, 79, 88, 90, 97, 113
contaminant, 12
contamination, 22
cost, 8, 21, 27, 28, 33, 58, 60, 81, 113
cost accounting, 21
cotton, 74
crop insurance, 32, 65
crop production, 16, 19, 22, 27, 29, 68, 87, 90, 92, 95, 102, 103
crop residue, 31, 106
crops, 11, 12, 13, 14, 15, 16, 19, 21, 22, 23, 26, 28, 29, 31, 32, 39, 54, 61, 63, 65, 71, 74, 75, 77, 78, 79, 82, 83, 97, 98, 99, 100, 101, 102, 103
crowds, 79
CRP, 63, 66, 94, 101, 111
cultivation, 7, 16, 19, 20, 30, 33, 75
cycling, 4, 13, 16

D

data analysis, 97
data set, 88
database, 95, 101

Index

decision makers, 18, 29, 33, 34, 35, 38
decision-making process, 37
deficiencies, 25
deforestation, 66
degradation, 13, 29
Department of Agriculture, v, vii, 1, 2, 3,
6, 42, 44, 45, 46, 47, 48, 49, 50, 51,
55, 108, 109, 110, 111, 112
Department of Commerce, 52
Department of Energy, v, vii, 1, 2, 3, 6,
12, 45, 52, 58, 111
deposition, 19
depth, 13
developing countries, 66, 102
direct cost, 26
diseases, 75
distribution, 16, 17, 60, 88, 97, 100, 104
diversification, 24
diversity, 7, 23, 33, 36
domestic demand, 81
domestic policy, 53, 107
drought, 20, 65, 75

E

Eastern Europe, 97
ecology, 3, 12, 24, 26
economic activity, 68
economic development, 9, 11, 38
economic growth, 71, 81
economics, 7, 8, 10, 27, 29, 32
ecosystem, 3, 4, 7, 8, 11, 12, 22, 23, 24,
29, 33
education, 2, 5, 38, 41
educators, 39
Egypt, 89
electricity, 31, 57
embargo, 57
emission, 58, 92, 94
employment, 40
encouragement, 33
endowments, 36, 80
energy, 2, 8, 9, 11, 14, 18, 21, 27, 30, 31,
32, 33, 34, 38, 39, 40, 57, 60, 61,

64, 65, 66, 68, 81, 82, 85, 86, 88,
94, 95, 103, 106, 108, 113
energy efficiency, 81
Energy Independence and Security Act,
10, 11, 53, 55, 111
energy input, 31
Energy Policy Act of 2005, 10, 58, 112
energy prices, 81
energy supply, 21
enforcement, 71
environment, 2, 4, 8, 14, 67, 68
environmental change, 18
environmental conditions, 18, 74
environmental effects, 29
environmental factors, 17, 18
environmental impact, 56, 57, 61, 66, 89
environmental issues, 25
environmental protection, 34
Environmental Protection Agency, 30,
46, 48, 52, 111, 112
environmental quality, 29, 56
environmental sustainability, 3, 12, 40
EPA, 60, 83, 84, 90, 91, 92, 93, 95, 96,
97, 98, 104, 106, 111, 112, 113
equilibrium, 68, 87, 97, 98, 106
equipment, 34, 37, 39
erosion, 8, 13, 16, 19, 20, 22, 74
ethanol, 2, 10, 11, 27, 28, 34, 35, 56, 57,
58, 59, 60, 61, 63, 76, 78, 82, 83,
84, 85, 86, 87, 88, 89, 90, 91, 92,
93, 95, 96, 97, 98, 101, 102, 113
EU, 55, 85, 99, 100, 102, 103
Europe, 36
European Commission, 102, 108
evaporation, 21
evapotranspiration, 21
evidence, 67, 74
evolution, 58, 68, 72, 79, 82
exporters, 78
exports, 19, 52, 53, 63, 78, 80, 87, 89,
99, 100

F

families, 2
Farm Bill, 10
farmers, 32, 34, 35, 37, 39, 66, 74, 75, 77
farms, 5, 17, 24, 36, 40
fertility, 4, 8, 15, 16, 22
fertilizers, 14, 16, 17, 20, 27, 65
fiber, viii, 2, 3, 7, 22, 24, 27, 52, 54, 66
field trials, 22
financial, 37, 38, 39
fire suppression, 15
flex, 27, 34
flexibility, 23, 32, 60, 78, 104
food, viii, 2, 3, 5, 7, 22, 24, 27, 35, 52, 54, 61, 66, 77, 78, 79, 90, 97, 98, 100, 101, 103, 104, 110
food production, 35, 61, 79
food products, 78, 100
food safety, 77
Ford, 47, 57
forecasting, 71
forest ecosystem, 17
forest management, 8, 20, 64
forest resources, 10
formation, 101
fuel prices, 28, 81

G

GDP, 68, 101
gene transfer, 24
genetics, 21
genomics, 4, 12, 16
geography, 36
global climate change, 9, 38
global competition, 68
global demand, 66
global markets, 64, 66, 87
global scale, vii, viii, 3, 8, 25, 40
global trade, 56, 102
globalization, 8
glycerin, 94

goods and services, 88
government policy, 28
graph, 59, 75
grasses, 19, 26, 60, 61, 74
grasslands, 11, 23, 31, 82
grazing, 88
greenhouse, viii, 3, 4, 11, 12, 13, 14, 15, 16, 18, 22, 25, 29, 30, 31, 52, 54, 103, 112
greenhouse gases, 4, 13, 14, 22
groundwater, 11
growth, 4, 8, 11, 14, 19, 20, 22, 25, 32, 37, 58, 68, 71, 75, 94, 101, 105
growth rate, 8, 94

H

habitat, 8, 13, 15, 22, 23, 29, 35
habitats, 4
harvesting, 7, 8, 22, 24, 75
health, 12, 15, 57
Henry Ford, 57
herbicide, 74
heterogeneity, 27
historical data, 21
history, 20, 55, 72, 77
hormones, 14
House, viii, 25, 42, 52, 53, 55
House Report, viii, 52, 53
housing, 64
human, 5, 8, 15, 18, 25, 28, 29, 36, 39, 41
human behavior, 5, 8, 25, 28, 29
human capital, 36, 41
human health, 15
hypoxia, 7, 35

I

ideal, 68
identification, 8
image, 2
imagery, 91
images, 2

Index 119

imports, 30, 57, 80, 100
improvements, 20, 32, 58, 74, 82, 88, 93, 99, 107
income, 36, 37, 64, 65, 81, 105
increased competition, 54
Independence, 49
India, 87, 89
indirect effect, 105
individuals, 37, 79
Indonesia, 81
industries, 34, 35, 36, 37, 38
industry, 11, 12, 20, 32, 36, 38, 39, 40, 41, 57, 58, 113
infrastructure, 7, 8, 17, 28, 31, 34, 36, 37, 38, 39, 40, 61, 80
insects, 22
institutions, 80
integration, 5, 19, 53, 61, 71, 82, 92, 107
International Monetary Fund, 107
international relations, 80
international trade, 33, 55, 64
investment, 6, 32, 37, 38, 39, 60
investments, 32, 37, 38, 41, 64, 75, 81
Iowa, 19, 43, 45, 48, 49, 50, 108
irrigation, 20, 21, 30, 65, 74
issues, 5, 6, 10, 25, 29, 36, 39, 77, 88, 90, 105, 106

J

Jordan, 11, 42

L

labor market, 8
labor markets, 8
land tenure, 35, 39
landscape, 3, 12, 17, 18, 21, 23, 24, 26
landscapes, 4, 7, 11, 22, 23, 24, 25, 26, 31, 35, 66
Latin America, 99
lead, 8, 14, 21, 54, 58, 77, 79, 82, 83, 106
legislation, 30

life cycle, 10, 21, 33, 61
lifestyle changes, 34
lignin, 113
liquids, 60
livestock, 27, 35, 67, 68, 76, 77, 90, 94, 98, 104, 105, 106
local conditions, 5
localization, 8
logistics, 38

M

machinery, 57, 64
magnitude, 54, 63, 64, 73, 78, 81, 88, 89, 93, 99, 105, 113
majority, 28, 60
management, 4, 5, 7, 13, 14, 15, 16, 17, 18, 19, 20, 21, 22, 23, 24, 25, 26, 27, 31, 35, 37, 39, 40, 52, 54, 60, 64, 74
market penetration, 113
marketing, 77
Maryland, 6, 45, 46
mass, 23, 30
materials, 28
matter, iv, 14, 57, 63, 75, 79
meat, 64, 79
median, 105
methodology, 61, 90, 92, 93, 104
Mexico, 8, 89
microbial communities, 4, 7, 13, 16, 17
migration, 66
mission, vii, 2, 3
missions, 2
Mississippi River, 19, 43
Missouri, 49
models, 4, 5, 7, 17, 18, 19, 21, 22, 23, 24, 25, 26, 27, 28, 29, 30, 31, 32, 33, 40, 52, 53, 56, 61, 67, 68, 71, 73, 77, 79, 81, 82, 83, 88, 90, 91, 92, 97, 102, 103, 106, 107, 108
modifications, 68, 88, 101
multiplier, 32, 86, 89
multiplier effect, 32
mutation, 74

Index

N

national policy, 34, 38
National Research Council, 18, 43
national security, 38
native species, 22
natural disaster, 81
natural disasters, 81
natural gas, 113
natural resources, 7, 10, 12, 25, 33, 66
negative effects, 20
Netherlands, 68
neutral, viii, 40, 53, 56
next generation, 11, 41
nitrogen, 4, 11, 14, 15, 16, 17, 20, 74
nitrous oxide, 4, 13, 14, 15, 16, 17, 18
nutrient, 4, 12, 13, 16, 19, 20, 28, 56, 77
nutrients, 13, 14, 15, 19, 22, 30, 74

O

obstacles, 61, 71
Office of Management and Budget, 52
oil, vii, 3, 27, 30, 31, 57, 81
oilseed, 106
Oklahoma, 50
operations, 32
opportunities, 4, 6, 8, 9, 10, 32, 34, 35, 39, 61, 68
opportunity costs, 80
optimization, 68
organic matter, 11, 13, 14
outreach, 3, 6, 38, 39, 41
outreach programs, 41
ownership, 35, 36, 40

P

paradigm shift, 10
parallel, 90
parameter estimates, 105
participants, 6, 10
partition, 75, 106

pasture, 11, 56, 63, 71, 80, 94, 96, 97, 100, 101
pathways, 60, 72, 91, 94
peer review, 92, 93
percolation, 21
performance measurement, 8
permit, 36
pesticide, 19, 56
pests, 75
petroleum, 11, 15, 27, 32, 57, 94
Petroleum, 94
phosphorus, 11, 14, 20
photosynthesis, 14
plant growth, 13, 14
plants, 13, 14, 21, 60, 93, 95, 113
policy, 8, 10, 17, 25, 33, 34, 35, 36, 37, 38, 40, 52, 53, 54, 55, 61, 64, 67, 68, 73, 78, 83, 88, 95, 103, 104, 105, 107
policy makers, 8, 10, 35, 61, 67
policy making, 40
policy options, 34
political instability, 81
pollination, 23
pollutants, 15
pools, 13, 87, 101, 106
population, 53, 55, 64, 67, 81, 101, 105
population growth, 53, 55, 64, 81, 105
portfolio, 5, 38, 60, 66
potential benefits, 12, 60
poverty, 90
precedent, 105
precipitation, 13, 17, 21, 30
preparation, iv, 19, 20, 22, 75
preservation, 78
price changes, 30, 63
price elasticity, 78, 88, 98, 104, 113
price signals, 34, 73
principles, 56
private investment, 74
probability, 78
producers, 5, 10, 18, 32, 35, 36, 37, 41, 53, 67, 78, 79, 89
product attributes, 39
product market, 106

Index

121

production costs, 6, 64, 81, 82
professionals, 38, 39
profit, 32
profitability, 37, 58, 65, 73, 77
project, 73, 80, 89, 90
protected areas, 80
protection, 12, 15
public interest, 81
public policy, 32
publishing, 53
pyrolysis, 60

Q

quantification, 13, 23, 61, 104

R

rainfall, 18, 21
rainforest, 82
rangeland, 94
raw materials, 28, 58, 61
reactions, 5, 34, 40
recognition, 53, 107
recommendations, iv, 41, 104, 113
Reform, 112
regions of the world, 87
regulations, 102, 104, 105
regulatory requirements, 55
reliability, 8
renewable energy, 9, 81
renewable fuel, viii, 52, 53, 55, 58, 81,
    90, 114
Renewable Fuel Standard, 10, 58, 59,
    65, 91, 92, 93, 95, 96, 111, 112
Renewable Fuels Association, 59
rent, 37
replacement rate, 94
requirements, 4, 7, 10, 15, 20, 21, 30, 34,
    37, 41, 59, 60, 61, 73, 93, 95, 98,
    101, 103, 112
researchers, 61, 64, 83, 87, 88, 93, 98
reserves, 29
residues, 26

resilience, 17, 24
resistance, 13, 17
resolution, 4, 25, 93, 94, 104
resource availability, 37, 38, 80
resource management, 37
resources, 3, 4, 8, 13, 35, 37, 38, 55
response, viii, 17, 18, 29, 30, 34, 53, 54,
    56, 58, 64, 66, 74, 78, 79, 83, 88,
    89, 90, 93, 98, 99, 100, 102, 103,
    104, 106, 113
responsiveness, 28, 71, 88, 99, 113
restoration, 15
retirement, 78
revenue, 28, 32, 37
RFS, 10, 58, 60, 61, 84, 91, 93, 94, 95,
    96, 112, 114
risk, 15, 30, 32, 37, 38, 39, 65, 106
risk management, 37, 38, 39, 106
risks, 23, 24, 32, 34, 35, 36
root, 4, 13
roots, 11, 13, 14, 21
Royal Society, 44
rules, 91, 93, 94
runoff, 8, 19, 20, 21, 22
rural areas, 32
rural development, 33

S

safety, 32
savings, 92, 113
sawdust, 57
scaling, viii, 52, 54, 61
science, vii, 3, 5, 6, 8, 10, 11, 21, 28, 34,
    36, 38, 39, 40, 41, 53, 75, 104, 105,
    106, 107
scope, 30, 55, 61, 63, 68, 72, 85, 99, 104
Secretary of Agriculture, 55
security, 9, 61
sediment, 19, 22
sediments, 19, 22
sensitivity, 68, 88, 98, 105, 106
shape, 38
shock, 100
shrubland, 97

signals, 54, 78
simulation, 102
skewness, 100
smog, 57
social benefits, 35
social consequences, 26
social costs, 28, 29, 35
social sciences, 38
social structure, 5, 34
society, 4, 7, 8, 26, 28, 36, 37, 38
soil erosion, 13, 19
solution, 39
South Africa, 89
South America, 36
soybeans, 11, 74, 77, 87, 89
species, 19, 23
stability, 22
stakeholders, 5, 6, 34, 35, 36, 38
starch, 58, 59, 60, 90, 113
state, vii, viii, 3, 6, 7, 22, 41, 52, 53, 56, 105, 113
state laws, 113
states, 68, 104
storage, 13, 14, 35, 36, 68, 77
structure, 8, 13, 14, 16, 23, 28, 40, 71, 79, 95, 97, 104, 107
substitutes, 35, 113
substitution, 68, 100
sugarcane, 14, 27, 88, 94
sulfur, 15
supplier, 76
supply chain, 61
suppression, 12, 16
sustainability, vii, 2, 3, 5, 6, 8, 10, 14, 21, 24, 25, 29, 30, 35, 39, 55
switchgrass, 19, 23, 106

T

tar, 81
target, 7, 11, 25, 41
tax incentive, 67
taxes, 37
teams, 25, 41
techniques, 24

technological advancement, 64
technological advances, vii, 3
technological change, 7, 17, 30, 34
technologies, 8, 30, 31, 32, 38, 40, 55, 57, 60, 74, 81, 83, 95, 105, 106
technology, 2, 10, 25, 41, 58, 68, 75, 88, 89
technology gap, 25, 41
temperature, 13, 14, 17, 18
tenure, 28, 37
Thailand, 89
threshold level, 58
trade, 33, 40, 64, 68, 73, 79, 80, 83, 87, 88, 89, 97, 100, 102, 104, 106
trading partners, 97
traits, 75
trajectory, 75
transformation, 100
transparency, 53
transpiration, 21
transport, 35, 60, 77, 102
transport costs, 102
transportation, 26, 34, 35, 36, 61, 79, 80, 105, 111
treatment, 31, 105, 106
triggers, 88
tropical forests, 82, 88

U

uniform, 75
United, v, 1, 10, 11, 15, 19, 20, 27, 28, 31, 33, 42, 51, 52, 53, 54, 58, 59, 61, 65, 67, 71, 72, 75, 77, 78, 82, 83, 87, 88, 89, 97, 98, 99, 102, 109, 110
United States, v, 1, 10, 11, 15, 19, 20, 27, 28, 31, 33, 42, 51, 52, 53, 54, 58, 59, 61, 65, 67, 71, 72, 75, 77, 78, 82, 83, 87, 88, 89, 97, 98, 99, 102, 109, 110
updating, 104
urban, 11, 29, 37
urban areas, 11, 29
urbanization, 18